Discovering Algebra

An Investigative Approach

Condensed Lessons for Make-up Work

DISCOVERING MATHEMATICS™

Key Curriculum Press
Innovators in Mathematics Education

Teacher's Materials Project Editor: Joan Lewis

Editorial Assistants: Christopher David, Shannon Miller

Writer: Stacey Miceli

Production Editor: Jacqueline Gamble

Copy Editor: Margaret Moore

Editorial Production Manager: Deborah Cogan

Production Director: Diana Jean Parks

Production Coordinator: Ann Rothenbuhler

Text Designer: Jenny Somerville

Art Editor: Jason Luz

Composition, Technical Art, Prepress: Interactive Composition Corporation

Art and Design Coordinator: Caroline Ayres

Cover Designer: Jill Kongabel

Printer: Data Reproductions

Executive Editor: Casey FitzSimons

Publisher: Steven Rasmussen

Cover Photo Credits: Background image: Pat O'Hara/DRK Photo. Boat image: Marc Epstein/DRK Photo. All other images: Ken Karp Photography.

Key Curriculum Press
1150 65th Street
Emeryville, CA 94608
(510) 595-7000
editorial@keypress.com
http://www.keypress.com

Printed in the United States of America
10 9 8 7 6 5 4 3 2 1 05 04 03 02 01

ISBN 1-55953-475-3

Contents

Chapter 10

Chapter 11

Introduction

Many of the key ideas in *Discovering Algebra: An Investigative Approach* are revealed through group investigations done in class. For this reason, it can be difficult for a student to make up for a missed lesson simply by reading the student text. To help you keep absent students current on the new algebra concepts, two-page condensed lessons have been written as abbreviated versions of the student lessons.

Each condensed lesson contains a summary of the investigation, with all the important results and their implications. In addition, most condensed lessons include a worked-out example similar to the example in the student lesson. While there is no replacement for peer interaction or teacher guidance, reading a condensed lesson along with the lesson in the student text helps a student understand the activities and important concepts that he or she missed. Students can read through a condensed lesson alone or with a parent or friend.

These condensed lessons can be copied two-sided to save paper. You may want to keep a file of copies handy so that they are available when you need them. Students could borrow the pages and return them to you when they complete their assignment for that lesson.

CONDENSED LESSON

0.1 The Same yet Smaller

In this lesson you will

- apply a **recursive rule** to create a **fractal design**
- use **fraction** operations to calculate partial areas of fractal designs
- review the **order of operations**

Investigation: Connect the Dots

You can create a **fractal design** by repeatedly applying a **recursive rule** to change a figure. First, you apply the rule to the starting, or Stage 0, figure to create the Stage 1 figure. Then, you apply it to the Stage 1 figure to create the Stage 2 figure, to the Stage 2 figure to create a Stage 3 figure, and so on. Usually, parts of later stages of a fractal design look like Stage 1. This feature is called **self-similarity.**

Steps 1–5 The diagram on page 2 of your book shows Stages 0–3 of a fractal design created by using the rule "Connect the midpoints of the sides of each upward-pointing triangle." If you continued applying this rule forever, you would get a design called the **Sierpiński triangle.**

On the Connect the Dots worksheet, create Stage 4 of the fractal design by following the directions in Step 4. As you work, think about the patterns from one stage to the next. Your completed Stage 4 triangle should look like the figure on page 9 of your book.

Steps 6–10 Suppose the area of the Stage 0 triangle is 1.

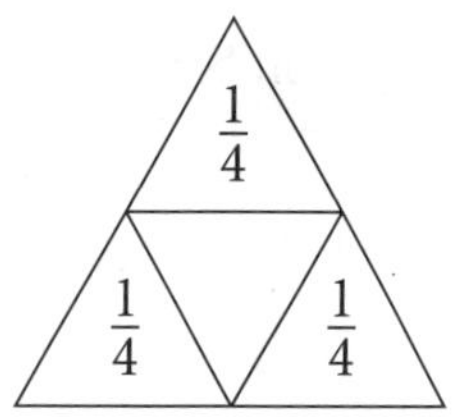

Stage 1

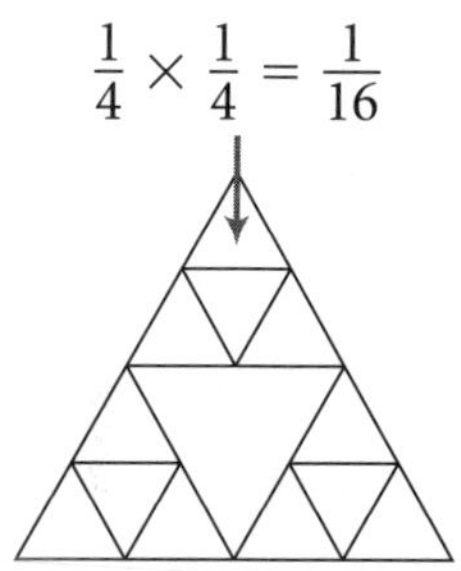

Stage 2

- The area of the smallest triangle at Stage 1 is $\frac{1}{4}$, and the combined area of the 3 upward-pointing triangles is $3 \times \frac{1}{4}$ or $\frac{3}{4}$.
- The area of the smallest triangle at Stage 2 is $\frac{1}{4} \times \frac{1}{4}$ or $\frac{1}{16}$. Because there are 9 smallest upward-pointing triangles, their combined area is $\frac{9}{16}$.
- At Stage 3, there are 27 (9×3) smallest upward-pointing triangles, each with area $\frac{1}{4} \times \frac{1}{16}$ or $\frac{1}{64}$. The combined area of these triangles is $\frac{27}{64}$.
- At Stage 4, there are 81 (27×3) smallest upward-pointing triangles, each with area $\frac{1}{4} \times \frac{1}{64}$ or $\frac{1}{256}$. The combined area is $\frac{81}{256}$.

Steps 11–12 The combined areas of triangles at different stages can be found. Here is an example.

- Suppose the area of the Stage 0 triangle is 8. Then the combined area of one smallest triangle at Stage 1, three at Stage 2, and two at Stage 3 is

$$\left(8 \times \frac{1}{4}\right) + \left(8 \times \frac{3}{16}\right) + \left(8 \times \frac{2}{64}\right) = 2 + \frac{3}{2} + \frac{1}{4} = 2 + \frac{6}{4} + \frac{1}{4} = 2 + \frac{7}{4}$$
$$= 2 + \frac{4}{4} + \frac{3}{4} = 3\frac{3}{4}$$

(continued)

You might want to check your work on a calculator. **Calculator Note 0A** shows you how to set your graphing calculator to give answers in fraction form.

When you evaluate an expression, be sure to follow the **order of operations:** simplify all expressions inside parentheses, *then* evaluate all powers, *then* multiply and divide from left to right, *then* add and subtract from left to right.

EXAMPLE

The diagram on page 5 of your book shows three stages of a different fractal design.

a. If the area of the Stage 0 triangle is 1, what is the combined area of one smallest triangle at Stage 1 and four smallest triangles at Stage 2?

b. If the area of the Stage 0 triangle is 18, what is the combined area of the shaded triangles?

▸ Solution

a. The area of the smallest triangle at Stage 1 is $\frac{1}{9}$.
The area of the smallest triangle at Stage 2 is $\frac{1}{9} \times \frac{1}{9}$ or $\frac{1}{81}$.
Now find the combined area of one smallest Stage 1 triangle $\left(\frac{1}{9}\right)$ and four smallest Stage 2 triangles $\left(4 \times \frac{1}{81}\right)$.

$$\left(\frac{1}{9}\right) + \left(4 \times \frac{1}{81}\right) = \frac{1}{9} + \frac{4}{81}$$ Multiply by 4 before you do any adding.

$$= \frac{9}{81} + \frac{4}{81}$$ Rewrite the fractions with a common denominator.

$$= \frac{13}{81}$$ Add the numerators.

The combined area is $\frac{13}{81}$.

b. The area of the Stage 0 triangle is 18, so you can find the combined area of the shaded triangles by multiplying each area by 18.

$$\left(\frac{1}{9} \times 18\right) + \left(\frac{4}{81} \times 18\right) = \frac{18}{9} + \frac{8}{9} = \frac{26}{9} = 2\frac{8}{9}$$

Or you can multiply the result from part a by 18.

$$\frac{13}{81} \times 18 = \frac{26}{9}$$

The combined area is $2\frac{8}{9}$.

CONDENSED LESSON

0.2 More and More

In this lesson you will

- look for patterns in **the way a fractal grows**
- use patterns to **make predictions**
- use **exponents** to represent repeated multiplication

Investigation: How Many?

The number of new upward-pointing triangles grows with each stage of the Sierpiński design. If you look closely, you will see a pattern you can use to predict the number of new triangles at any stage.

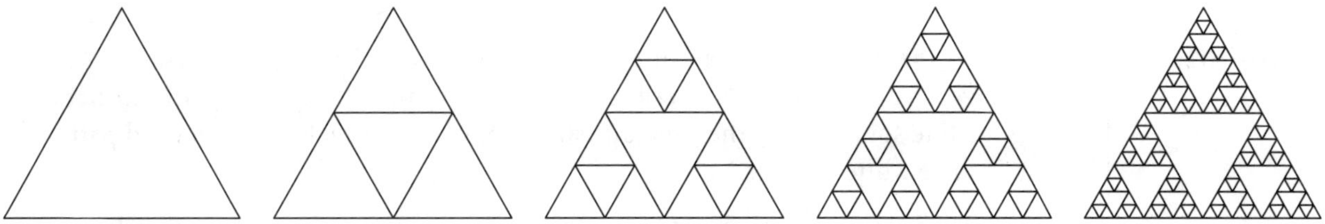

Stage 0 Stage 1 Stage 2 Stage 3 Stage 4

If you count the new upward-pointing triangles at Stages 0–4, you will get the results shown in this table. Notice that the number of new upward-pointing triangles at each stage is 3 times the number in the previous stage.

Stage number	Number of new upward-pointing triangles	
0	1	
1	3	(3 is 3 times 1.)
2	9	(9 is 3 times 3.)
3	27	(27 is 3 times 9.)
4	81	(81 is 3 times 27.)

Continuing this pattern, the number of new triangles at Stage 5 is $3 \cdot 81$ or 243, the number at Stage 6 is $3 \cdot 243$ or 729, and the number at Stage 7 is $3 \cdot 729$ or 2187.

To find the number of new upward-pointing triangles at Stage 15, you can continue multiplying by 3 to find the number at Stage 8, Stage 9, Stage 10, and so on up to Stage 15. Or you can use this pattern:

Number at Stage 1 $= 3$ (one 3)

Number at Stage 2 $= 3 \cdot 3$ (product of two 3's)

Number at Stage 3 $= 3 \cdot 3 \cdot 3$ (product of three 3's)

Number at Stage 4 $= 3 \cdot 3 \cdot 3 \cdot 3$ (product of four 3's)

$\vdots$

(continued)

Continuing this pattern, the number of new triangles at Stage 15 is the product of fifteen 3's: $3 \cdot 3 \cdot 3 \cdot 3 \cdot 3 \cdot 3 \cdot 3 \cdot 3 \cdot 3 \cdot 3 \cdot 3 \cdot 3 \cdot 3 \cdot 3 \cdot 3 = 14{,}348{,}907$.

EXAMPLE Describe how the number of white squares is growing in this fractal design.

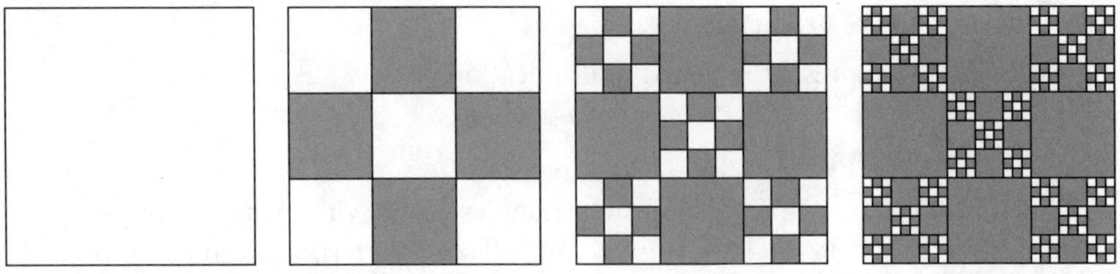

Stage 0 Stage 1 Stage 2 Stage 3

Solution Study how the design changes from one stage to the next. The recursive rule is "At each stage, create a 3-by-3 checkerboard pattern (with white squares in the corners) in each white square from the previous stage." Notice that each checkerboard pattern contains five white squares.

Stage 1 has five white squares. At Stage 2, five white squares are created in each of the white squares from Stage 1, for a total of $5 \cdot 5$ or 25 white squares. At Stage 3, five white squares are created in each of the 25 squares from Stage 2, for a total of $5 \cdot 25$ or 125 white squares.

You can show these results in a table.

	Number of white squares		
Stage number	**Total**	**Repeated multiplication**	**Exponent form**
1	5	5	5^1
2	25	$5 \cdot 5$	5^2
3	125	$5 \cdot 25$ or $5 \cdot 5 \cdot 5$	5^3

In the last column, the small, raised numbers, called **exponents,** tell you how many 5's are multiplied together.

Notice that, for each stage, the exponent is equal to the stage number. You can use this idea to find the number of white squares at any stage. For example, the number of white squares at Stage 4 is 5^4 or 625, and the number at Stage 10 is 5^{10} or 9,765,625. To see if this pattern works for Stage 0, enter 5^0 into your calculator. (See **Calculator Note 0B** to learn how to enter exponents.) The result is 1, which is the number of white squares at Stage 0. This fits the pattern.

CONDENSED LESSON

0.3 Shorter yet Longer

In this lesson you will

- find a **recursive rule** for generating a fractal design
- use **fractions** to express the lengths of segments in various stages of the fractal design
- use **exponents** to describe a rule for finding the total length of any stage of the fractal design
- use a calculator to **convert fractions to decimals**

Investigation: How Long Is This Fractal?

The *Koch curve* is a fractal created from line segments. Think about how the "curve" changes from one stage to the next.

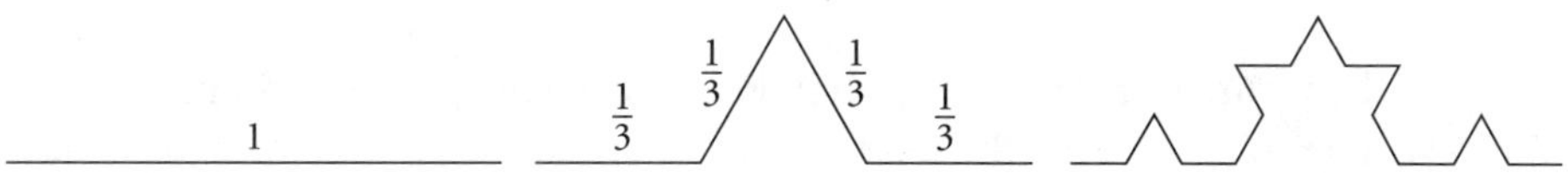

To discover the recursive rule for creating the design, study what happens from Stage 0 to Stage 1.

One possible recursive rule is "To get the next stage, divide each segment from the previous stage into thirds and build a equilateral triangle above the middle third. Then, remove the 'bottom' of each triangle."

You can make a table to show how the "curve" changes from stage to stage. Notice that the number of segments at each stage is 4 times the number in the previous stage and that the length of each segment is $\frac{1}{3}$ the length of the segments at the previous stage.

			Total length (Number of segments times length of segments)	
Stage number	**Number of segments**	**Length of each segment**	**Fraction form**	**Decimal form (rounded)**
0	1	1	1	1
1	4	$\frac{1}{3}$	$4 \cdot \frac{1}{3} = \frac{4}{3}$	1.3
2	$4 \cdot 4 = 4^2 = 16$	$\frac{1}{3} \cdot \frac{1}{3} = \left(\frac{1}{3}\right)^2 = \frac{1}{9}$	$4^2 \cdot \left(\frac{1}{3}\right)^2 = \left(\frac{4}{3}\right)^2 = \frac{16}{9}$	1.8

If you continue the patterns in the table, you'll see that Stage 3 has $4 \cdot 4 \cdot 4$ or 4^3 segments and that the length of each segment is $\frac{1}{3} \cdot \frac{1}{3} \cdot \frac{1}{3}$ or $\left(\frac{1}{3}\right)^3$. So the total length of the Stage 3 figure is $4^3 \cdot \left(\frac{1}{3}\right)^3$, which can be written $\left(\frac{4}{3}\right)^3$. This simplifies to $\frac{64}{27}$, or about 2.37.

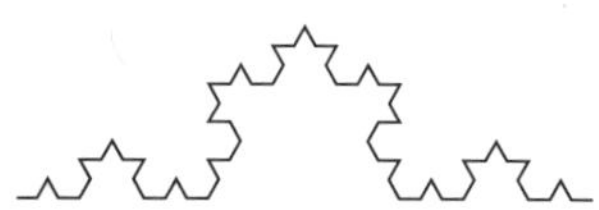
Stage 3

(continued)

Notice that, when you write the total length of a stage using exponents, the exponent is equal to the stage number. So the total length of the Stage 4 figure is $\left(\frac{4}{3}\right)^4$, which simplifies to $\frac{256}{81}$, or about 3.16.

EXAMPLE Look at the beginning stages of this fractal:

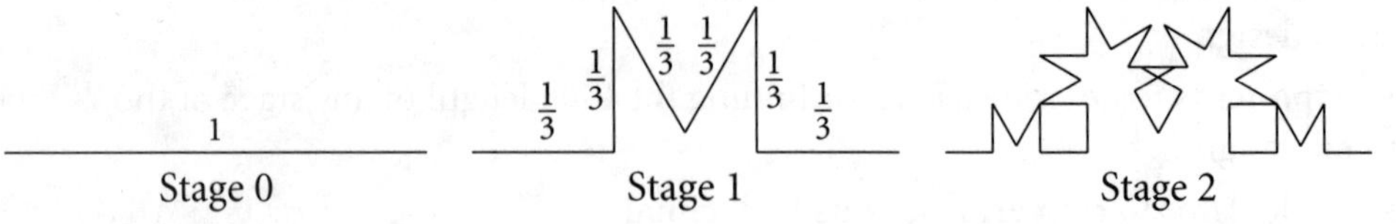

a. Describe the fractal's recursive rule.

b. Find its length at Stage 2.

c. Write an expression for its length at Stage 23.

▶ Solution

a. To find the recursive rule, consider how the fractal changes from Stage 0 to Stage 1. The rule is "To get the next stage, divide each segment of the previous stage into thirds. Then, replace the middle third with an M made from segments that are each $\frac{1}{3}$ the length of the segments from the previous stage."

b. To find the length of the fractal at Stage 2, first look at its length at Stage 1. The Stage 1 figure has six segments of length $\frac{1}{3}$. Each Stage 2 segment is replaced by six new segments, so the Stage 2 figure has $6 \cdot 6$ or 6^2 segments. Each Stage 2 segment is $\frac{1}{3}$ the length of the Stage 1 segments, so each Stage 2 segment has length $\frac{1}{3} \cdot \frac{1}{3}$ or $\left(\frac{1}{3}\right)^2$. The total length at Stage 2 is $6^2 \cdot \left(\frac{1}{3}\right)^2$, which can be written as $\left(\frac{6}{3}\right)^2$ or 2^2. So the total length is 4.

c. At each stage, each segment from the previous stage is replaced with six new segments. The length of each new segment is $\frac{1}{3}$ the length of the segment at the previous stage. By Stage 23, this has been done 23 times. The Stage 23 figure is $6^{23} \cdot \left(\frac{1}{3}\right)^{23}$ or $\left(\frac{6}{3}\right)^{23}$ or 2^{23} long.

CONDENSED LESSON

0.4 Going Somewhere?

In this lesson you will

- review operations with **integers**
- use a recursive process to **evaluate expressions**
- use a calculator to evaluate expressions
- identify the **attractors** of expressions

Investigation: A Strange Attraction

You have seen how recursion can be used to create fractal designs. In this investigation you look at recursive processes involving numerical expressions.

Steps 1–5 For example, consider this expression:

$-2 \cdot \square + 3$

Start with any number, put it in the box, and do the computation. For example, here's the result if you start with 0:

$-2 \cdot (0) + 3 = 0 + 3 = 3$

Now, take the result, 3, and put it in the box in the original expression.

$-2 \cdot (3) + 3 = -6 + 3 = -3$

You can continue this process, each time using your result from the previous stage. Here are the results for the first 10 stages.

Original expression: $-2 \cdot \square + 3$ Starting number (at Stage 0): 0		
Stage number	**Expression**	**Result**
1	$-2 \cdot 0 + 3$	3
2	$-2 \cdot 3 + 3$	-3
3	$-2 \cdot -3 + 3$	9
4	$-2 \cdot 9 + 3$	-15
5	$-2 \cdot -15 + 3$	33
6	$-2 \cdot 33 + 3$	-63
7	$-2 \cdot -63 + 3$	129
8	$-2 \cdot 129 + 3$	-255
9	$-2 \cdot -255 + 3$	513
10	$-2 \cdot 513 + 3$	-1023

Steps 6–9 You can draw a number line diagram to show how the value of the expression changes at each stage. This diagram shows the results for Stages 0–5.

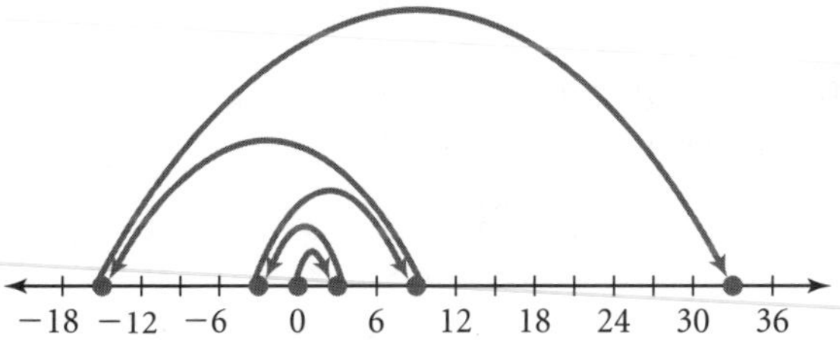

Now, try evaluating $-2 \cdot \square + 3$ recursively for a different starting value. When you evaluate $-2 \cdot \square + 3$ recursively, the results get farther and farther apart. For some expressions, no matter what value you start with, the results get closer and closer to a particular number.

(continued)

EXAMPLE

What happens when you evaluate this expression recursively with different starting numbers?

$0.5 \cdot \square - 3$

▶ ***Solution***

Let's see what happens for a couple different starting numbers.

In both cases, the results seem to get closer to one number, perhaps −6. If, no matter what starting value you try, the results get closer and closer to −6, then −6 is an **attractor** for the expression.

To check whether −6 is an attractor, use it as the starting number. If the result stays at −6, then it is an attractor.

$0.5 \cdot (-6) - 3 = -3 - 3 = -6$

Because you get back exactly what you started with, −6 is an attractor for the expression $0.5 \cdot \square - 3$.

Starting number: 4

$0.5 \cdot (4) - 3 = 2 - 3 = -1$

$0.5 \cdot (-1) - 3 = -0.5 - 3 = -3.5$

$0.5 \cdot (-3.5) - 3 = -1.75 - 3 = -4.75$

$0.5 \cdot (-4.75) - 3 = -2.375 - 3 = -5.375$

$0.5 \cdot (-5.375) - 3 = -2.6875 - 3 = -5.6875$

$0.5 \cdot (-5.6875) - 3 = -2.84375 - 3 = -5.84375$

Starting number: −10

$0.5 \cdot (-10) - 3 = -5 - 3 = -8$

$0.5 \cdot (-8) - 3 = -4 - 3 = -7$

$0.5 \cdot (-7) - 3 = -3.5 - 3 = -6.5$

$0.5 \cdot (-6.5) - 3 = -3.25 - 3 = -6.25$

$0.5 \cdot (-6.25) - 3 = -3.125 - 3 = -6.125$

$0.5 \cdot (-6.125) - 3 = -3.0625 - 3 = -6.0625$

As you have seen, not all expressions have an attractor. Other expressions have attractors that are difficult or impossible to find. With practice, you may be able to predict the attractors for some simple expressions without actually doing any computations.

CONDENSED LESSON

0.5 Out of Chaos

In this lesson you will

- review measuring distances and finding fractions of distances
- use a chaotic process to create a pattern by hand
- use a chaotic process to create a pattern on a calculator

Investigation: A Chaotic Pattern?

When you roll a die, the results are **random.** If you looked at the results of many rolls, you would not expect to see a pattern that would let you predict exactly when and how often a particular number will appear. However, as you will see, sometimes a random process, like rolling a die, can be used to generate an orderly result.

Steps 1–9 Suppose you start with this triangle and choose a starting point anywhere inside the triangle.

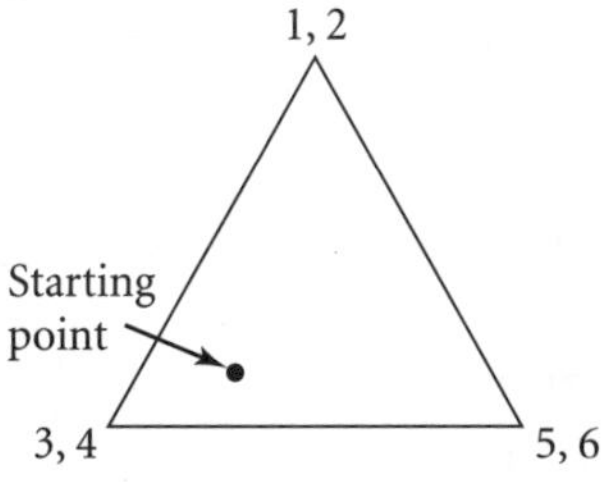

Then you roll a die and mark the point halfway from the starting point to the **vertex** labeled with the number on the die. For example, if you roll a 2, you would mark the point halfway between the starting point and the top vertex. This point becomes the starting point for the next stage.

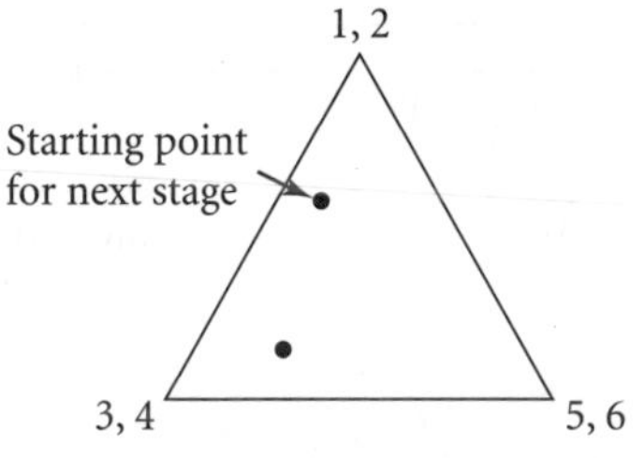

Roll the die again and mark the point halfway between the new starting point and the vertex labeled with the number on the die.

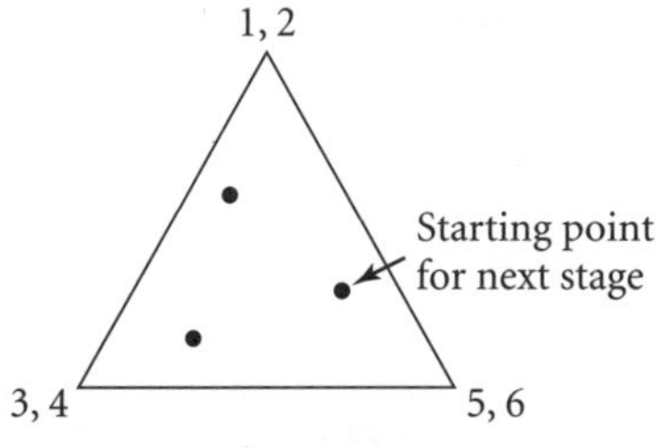

(continued)

If you repeat this process many times, an interesting pattern will emerge.

Rolling the die and plotting the points by hand is time-consuming. Fortunately, your calculator can do the same process much more quickly. The Chaos program chooses a starting point and follows the procedure described above to plot 1000 points.

Steps 10–12 Try running the program on your calculator. It will take a while for the calculator to plot all the points, so be patient.

When the program is finished, you should see this familiar pattern on the screen.

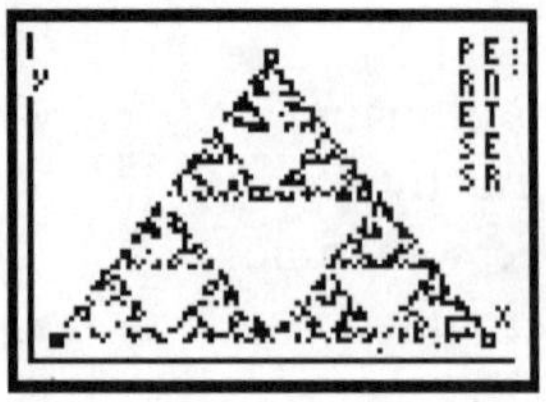

The pattern of points looks like the Sierpiński triangle!

Mathematicians use the word *chaotic* to describe an orderly procedure that produces a random-looking result. Here we started with a random procedure that produces an ordered-looking result. The orderly result is called a *strange attractor.* No matter what starting point you choose, the points will "fall" toward the shape. Many fractal designs, like the Sierpiński triangle, are strange attractors. Accurate measurements are essential to seeing a strange attractor form.

EXAMPLE

Find point F three-fifths of the way from D to E. Give the distance from D to F in centimeters.

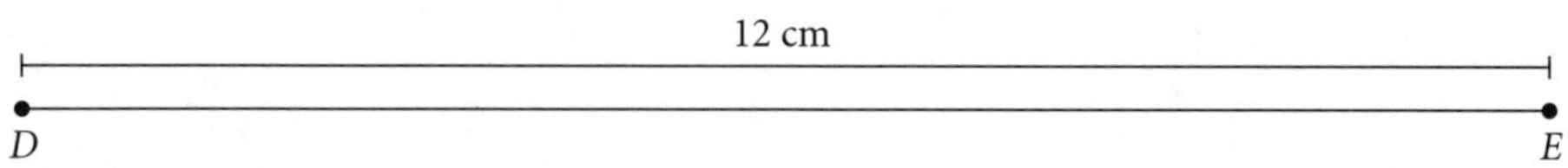

Have your ruler handy so that you can check the measurements. Use your calculator to check the computations.

▶ ***Solution***

Measuring segment DE shows that it is about 12 cm long. Three-fifths of 12 is $\frac{3}{5} \cdot 12$, which you can rewrite as $\frac{36}{5}$ or 7.2. Place point F 7.2 cm from point D.

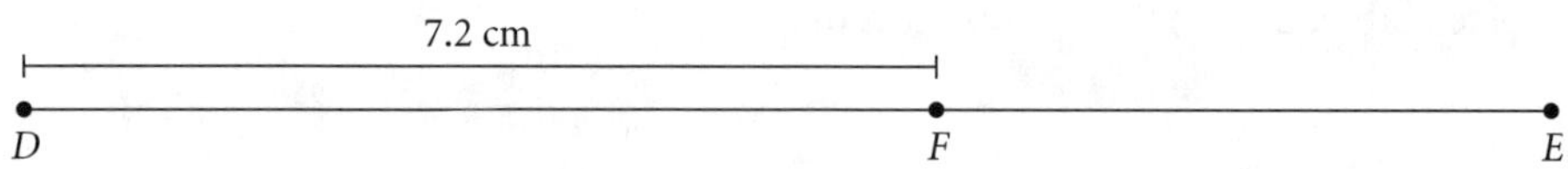

CONDENSED
LESSON

1.1 Bar Graphs and Dot Plots

In this lesson you will

- **interpret and create** a variety of graphs
- find some **summary values** for a data set
- **draw conclusions** about a data set based on graphs and summary values

This **pictograph** shows the number of pets of various types that were treated at Uptown Animal Hospital in one week.

Specific information, such as the number of pets of each type, is called **data.** You can often display data in tables and graphs.

EXAMPLE | Create a table of data and a **bar graph** from the pictograph above.

▸ ***Solution***

This table shows the number of pets of each type. Remember that each symbol on the pictograph represents two pets.

Pets Treated in One Week

Dogs	Cats	Ferrets	Birds	Hamsters	Lizards
17	12	8	5	4	2

The bar graph shows the same data. The height of a bar shows the total in that **category,** in this case, a particular type of pet. You can use the *scale* on the *vertical axis* to measure the height of each bar.

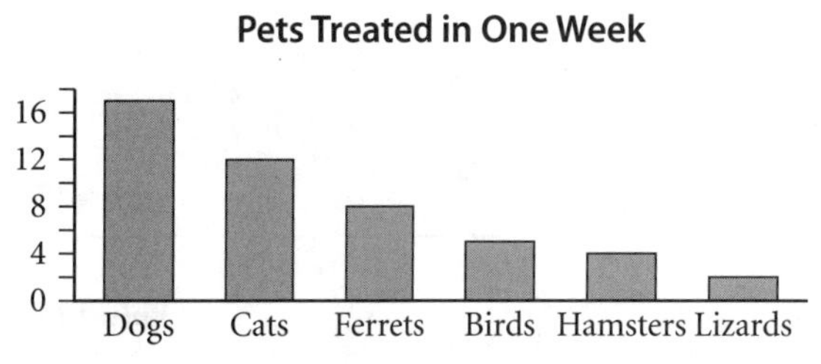

Bar graphs gather data into categories, allowing you to quickly compare values for each category.

A **dot plot** shows each item of *numerical* data above a *number line,* or *horizontal axis.* Dot plots make it easy to see gaps and clusters in a data set, as well as how the data **spreads** along the axis.

Investigation: Picturing Pulse Rates

This investigation uses data about pulse rates. Pulse rates vary, but a healthy person at rest usually has a pulse rate between certain values. A person with a pulse rate that is too fast or too slow may need medical attention.

(continued)

This data set gives pulse rates, in beats per minute, for a group of 30 students.

68	60	76	68	64	80	72	76	92	68	56	72	68	60	84
72	56	88	76	80	68	80	84	64	80	72	64	68	76	72

For this data set, the **minimum** (lowest) value is 56 and the **maximum** (highest) value is 92. The minimum and maximum describe the spread of the data. For example, you could say, "The pulse rates are between 56 and 92 bpm." Based only on this data, it appears that a pulse rate of 80 bpm would be "normal," while a pulse rate of 36 bpm would be too low.

To make a dot plot of the pulse rates, first draw a number line with the minimum value, 56, at the left end. Select a scale and label equal **intervals** until you reach the maximum value, 92.

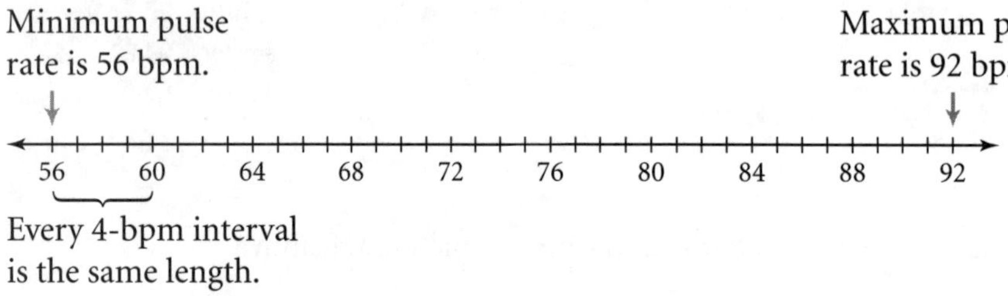

For each value in the data set, put a dot above that value on the number line. When a value occurs more than once, stack the dots. For example, the value 64 occurs three times in the data set, so there are three dots above 64. Be sure to label the axis so that it is clear what the data are.

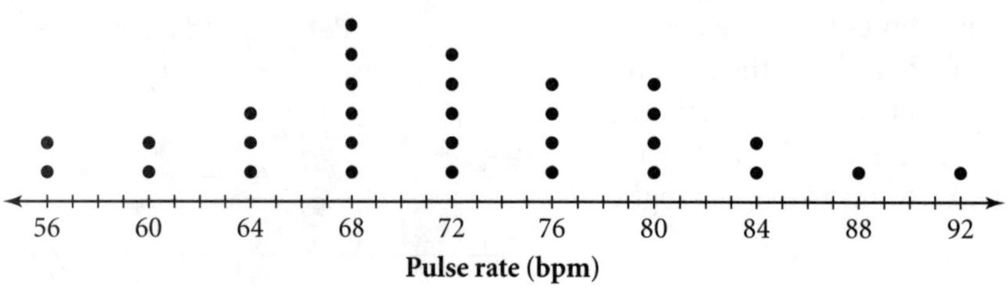

The **range** of a data set is the *difference* between the maximum and minimum values. For this data, the range is $92 - 56$ or 36 bpm. Note that the range does not tell you about the actual values in a data set. If a paramedic tells you that normal pulse rates have a range of 12, this would not tell you anything about the minimum or maximum value or any values in-between.

Looking at the dot plot, you can see that there are a few values near the maximum and minimum, but most values cluster between 64 and 80. The value 68 occurs most often, followed by 72. This information gives you an idea of what is a normal pulse for this class, but it could not be used to say what is a "normal" pulse for all people. Many factors, including age, affect pulse rates.

Statistics is a word we use many ways. It is often used to refer to numbers that describe or summarize data. For example, you could collect pulse-rate data from thousands of people and then determine a single value that can be considered "normal." This single value can also be called a statistic.

CONDENSED LESSON 1.2

Summarizing Data with Measures of Center

In this lesson you will

- find **measures of center** for a data set
- choose the **most meaningful measure of center** for a situation
- look at the influence of **outliers** on the **mean** of a data set

Read the statements at the beginning of the lesson. Each statement uses a single number—called a **measure of center**—to describe what is typical about a set of data. The first statement uses the **mean,** or average. The second statement uses the **median,** or middle value. The third statement uses the **mode,** or most frequent value.

Investigation: Making "Cents" of the Center

You can use this collection of pennies to help you understand the mean, median, and mode.

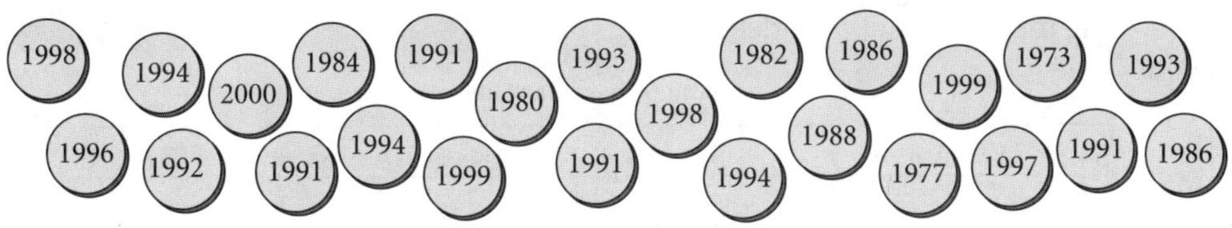

To find the median of the mint years, line up the pennies from oldest to newest.

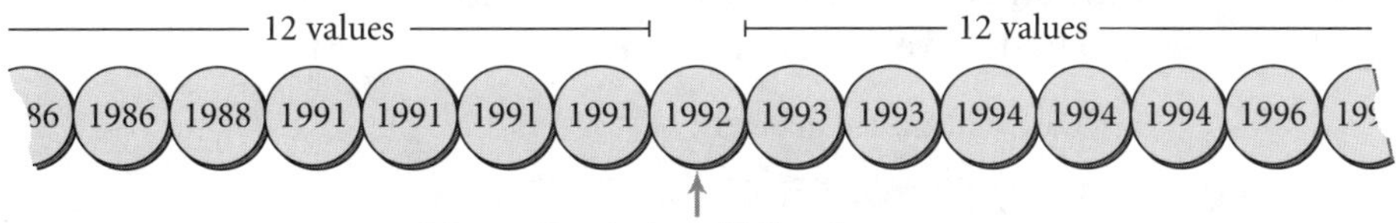

The median is the middle value, 1992.

To find the mode, stack the pennies with the same mint year.

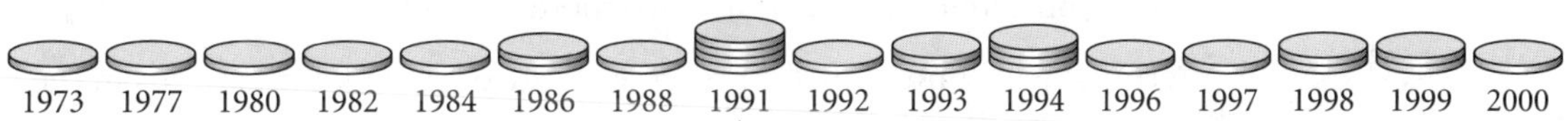

The mode is the year with the tallest stack, 1991.

To find the mean, find the sum of the mint years for all the pennies and divide by the number of pennies.

$$\text{mean} = \frac{\textit{sum of mint years}}{\textit{number of pennies}} = \frac{49{,}767}{25} = 1990.68.$$ You might round the mean up to 1991.

Here is a dot plot of the mint years with the median, mode, and mean labeled.

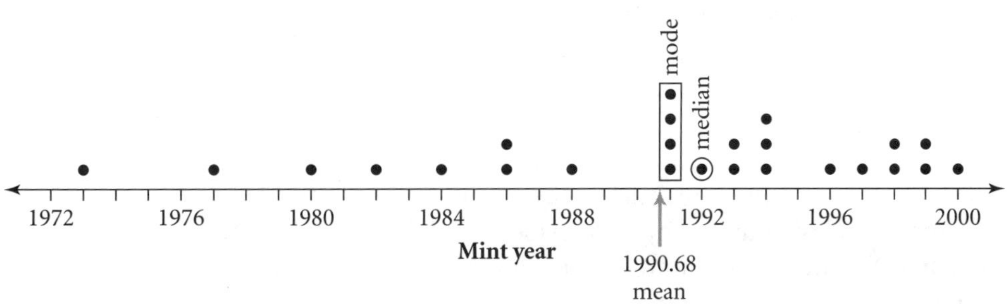

(continued)

Now enter the penny data into a calculator list, and use your calculator to find the mean and median. (See **Calculator Notes 1A and 1B.** Refer to **Calculator Note 1J** to check the settings on your calculator.)

Read the descriptions of the measures of center on page 45 of your book. The mean and the median of a set of data may be quite different, and the mode, if it exists, may not be useful. You will need to decide which measure is most meaningful for a given situation.

EXAMPLE

This data set shows the number of videos rented at a video store each day for 14 days. {54, 75, 2, 68, 49, 53, 5, 63, 54, 70, 65, 68, 71, 60}

a. Find the measures of center.

b. Which measure best represents the data?

Solution

a. The mean is about 54 videos.

$$\underbrace{\frac{\overbrace{54 + 75 + 2 + 68 + 49 + 53 + 5 + 63 + 54 + 70 + 65 + 68 + 71 + 60}^{\text{Sum of the data values}}}{14}}_{\text{Number of data values}} \approx \underset{\text{mean}}{54}$$

Because there are an even number of values, the median is the number halfway between the two middle values. In this case, the median is 61.5.

Data values listed in order

2, 5, 49, 53, 54, 54, 60, 63, 65, 68, 68, 70, 71, 75

The median is halfway between 60 and 63.

This data is **bimodal**—that is, it has two modes, 54 and 68.

b. To determine which measure best represents the data, look for patterns in the data and consider the shape of the graph.

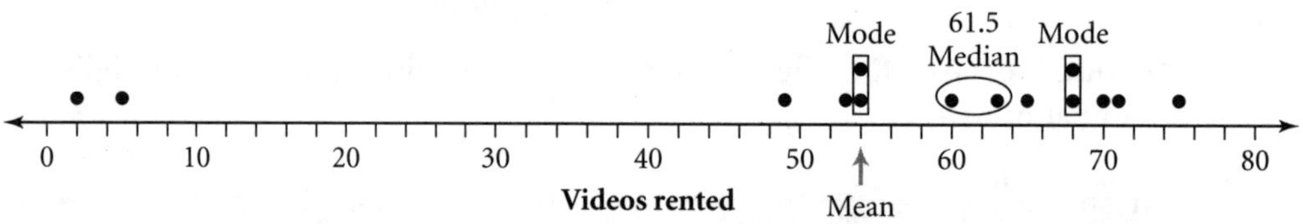

Except for two values (2 and 5), the data are clustered between 49 and 75. The values 2 and 5, called **outliers,** bring the mean, 54, too far to the left of most of the data values to be the best measure of center. One of the modes is the same as the mean. The other is too high. The median, 61.5, seems to summarize the data best.

In the example in your book, the outliers are much greater than the rest of the data. These values pull the mean up. Using the mean to describe data that includes outliers can be misleading. You need to be aware of these limitations when you read advertisements and reports that give measures of center.

Five-Number Summaries and Box Plots

In this lesson you will

- find **five-number summaries** for data sets
- interpret and create **box plots**
- draw conclusions about a data set based on graphs and summary values
- use a calculator to make a box plot

The table on page 50 of your book shows the total number of points scored by each player on the Chicago Bulls during the 1997–98 NBA season.

The **five-number summary** can give you a good picture of how the Bulls scored as a team. A five-number summary uses five boundary points: the minimum and the maximum, the median (which divides the data in half), the **first quartile** (the median of the first half), and the **third quartile** (the median of the second half).

The first quartile (Q1), the median, and the third quartile (Q3) divide the data into four equal groups. The example in your book illustrates how to find the five-number summary.

A **box plot,** or **box-and-whisker plot,** is a visual way to show the five-number summary. This box plot shows the spread of the Bulls' scoring data. Notice how the five-number summary is shown in the plot.

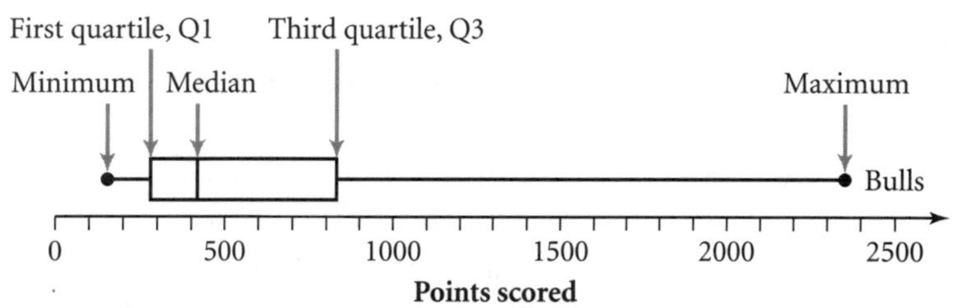

Even though the four sections of the plot—the two whiskers and the two halves of the box—are different lengths, each represents $\frac{1}{4}$ of the data. So, for example, the long, right whisker represents the same number of data items as the short, left whisker. Notice that most of the data values are concentrated in the left part of the graph. The long whisker that takes up most of the plot represents only $\frac{1}{4}$ of the values (including Michael Jordan's).

Investigation: Pennies in a Box

In this investigation, you make a box plot of the penny data from Lesson 1.2.

Steps 1–6 First you need to list the values in order and find the five-number summary. Here is some sample data with the minimum, maximum, median, and quartiles labeled. You may want to follow the steps with your data from Lesson 1.2.

Minimum ↓ (1973); The first quartile is 1986, the median of the first half. ↓

First half of data → 1973 1977 1980 1982 1984 1986 1986 1988 1991 1991 1991 1991

Median → 1992

Second half of data → 1993 1993 1994 1994 1994 1996 1997 1998 1998 1999 1999 2000

↑ The third quartile is 1996.5, the median of the second half. ↑ Maximum (2000)

(continued)

To make a box plot, draw the horizontal axis (you can use the same scale you used for the dot plot in Lesson 1.2). Draw a short vertical segment just above the median value. Do the same for the first and third quartiles. Then, draw dots over the minimum and maximum values.

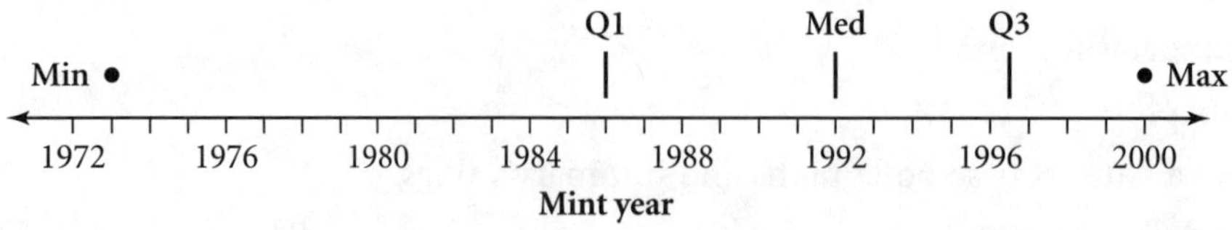

To finish the plot, draw a "box" with ends at the first and third quartiles, and draw "whiskers" extending from the ends of the box to the minimum and maximum values.

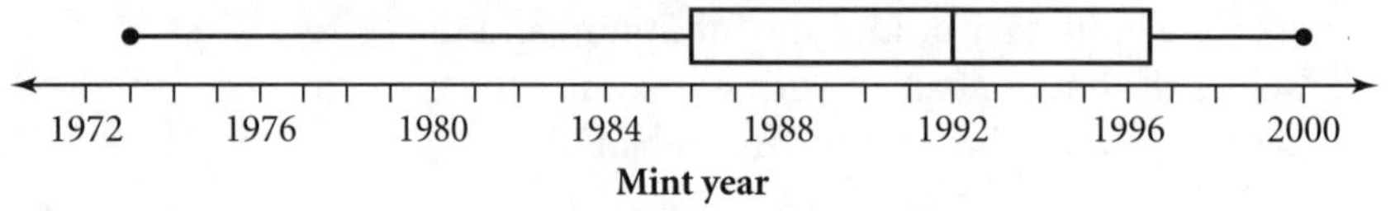

Compare the box plot with a dot plot of the same data.

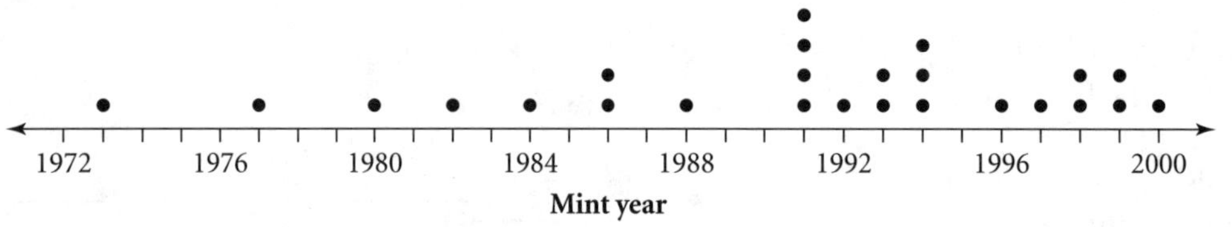

Both plots show that most of the values are greater than 1986. The box plot makes it easier to locate the five-number summary, but, unlike the dot plot, it does not show individual values or how many values there are. If seeing each data value is important, then a box plot is not the best way to display your data.

Remember that the four sections of a box plot—the two whiskers and the two halves of the box—each represent about the same number of data items. The box plot shows that $\frac{3}{4}$ of the mint years are between 1986 and 2000, while only $\frac{1}{4}$ are between 1973 and 1986.

Steps 7–9 Clear any old data from your calculator and enter the mint years into list L1. Draw a calculator box plot (see **Calculator Note 1C**). Compare your calculator plot to the one above. If you use the trace function, you can see the five-number summary for the data.

Step 10 The difference between the first quartile and the third quartile is the **interquartile range,** or **IQR**. Like the range, the interquartile range helps describe the spread in the data. For the sample penny data above, the range is 2000 − 1973 or 27 and the interquartile range is 1996.5 − 1986 or 10.5.

You can compare two data sets by making two box plots on the same axis. The box plots on page 52 of your book summarize the final test scores for two algebra classes. You can see that Class A has the greater range of scores and the greater IQR. In both classes, half of the students scored above 80. In Class B, all the students scored above 65, whereas in Class A, fewer than three-quarters of the students scored above 65.

CONDENSED LESSON 1.4

Histograms and Stem-and-Leaf Plots

In this lesson you will

- interpret and create **histograms**
- interpret **stem-and-leaf plots**
- choose appropriate **bin** widths for histograms
- use a calculator to make histograms

A dot plot includes a dot for each value in a data set. If a data set contains a large number of values, a dot plot might not be practical. A histogram is related to a dot plot but is more useful when you have a large set of data. Read the information about histograms on page 57 of your book. Notice how the bin width chosen for the histogram affects the appearance of the graph.

Investigation: Hand Spans

Your hand span is the distance from the tip of your thumb to the tip of your little finger when your fingers are spread.

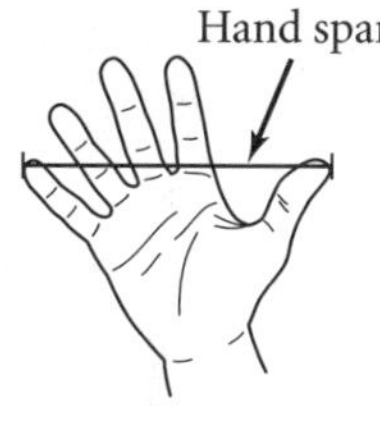

Steps 1–6 Here are hand spans, to the nearest half-centimeter, of students in one algebra class.

19	18.5	20.5	21.5	18.5	17.5	22	22.5	19.5	20	24	18
16.5	28	19	20	20.5	24	15	17	19	18	21	21

To make a histogram of this data, you first need to choose a bin width. As a general rule, try to choose an interval that will give you from 5 to 10 bins. For this data, the range is $28 - 15$ or 13. If you use a bin width of 2, you will have 7 bins.

Next, find the number of data values in each bin. These values are called **frequencies.** Note that a bin contains the left boundary value, but not the right. So the bin from 17 to 19 includes the values 17, 17.5, 18, and 18.5, but *not* 19. The value 19 belongs in the bin from 19 to 21.

Bin	15 to 17	17 to 19	19 to 21	21 to 23	23 to 25	25 to 27	27 to 29
Frequency	2	6	8	5	2	0	1

Now, draw the axes. Scale the horizontal axis to show hand-span values from 15 to 29 in intervals of 2. Scale the vertical axis to show frequency values from 0 to 8.

Finally, draw in bars to show the frequency values in your table.

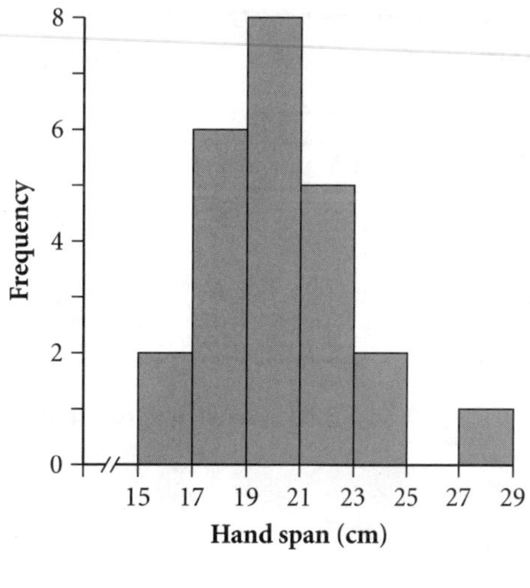

(continued)

Steps 7–9 If you enter the hand-span data into list L1 of your calculator, you can experiment with different bin widths. (See **Calculator Note 1D** for instructions on creating histograms.) Here are histograms of the same data using different bin widths.

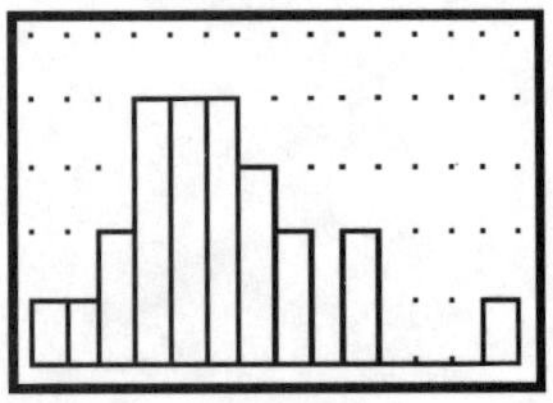

[15, 29, 1, 0, 5, 1]

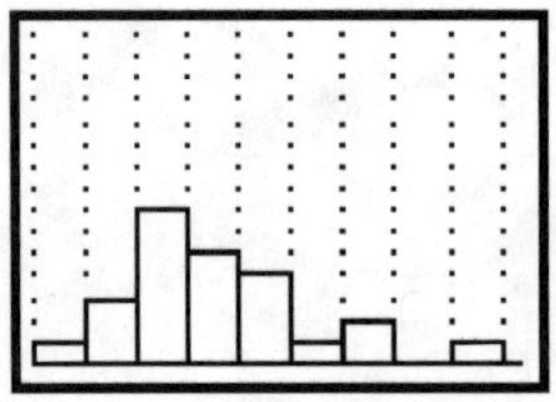

[15, 29, 1.5, 0, 15, 1]

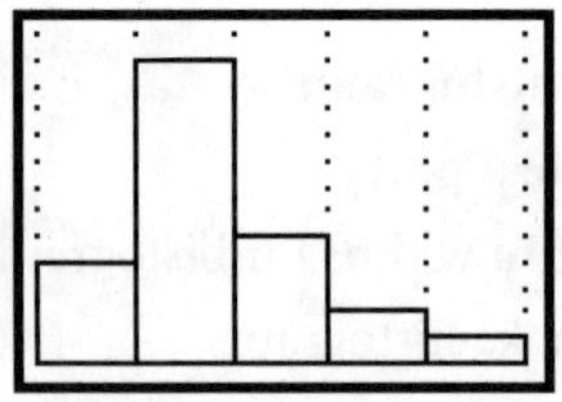

[15, 30, 3, 0, 15, 1]

For this data, a bin width of 1.5 or 2 works well. Both bin widths give a nice picture of how the values are distributed, indicating where data values are clustered and where there is a gap in the values. A bin width of 1 shows additional gaps but has a lot of bars. With a bin width of 3, there are too few bars to get a good picture of the distribution, and the gap on the right side is hidden.

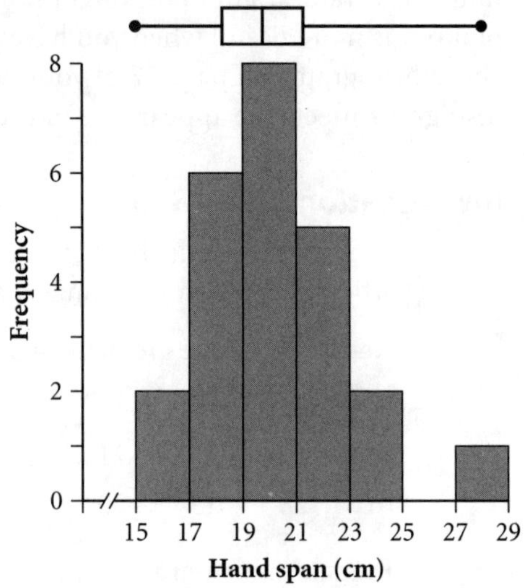

At right a box plot has been drawn on the same axes as the histogram. Unlike the histogram, the box plot does not show the gap in the data, and it does not give any indication of how many values are in each interval.

Like a histogram, a **stem plot,** or stem-and-leaf plot, groups data values into intervals. However, a stem plot shows each individual data item. For this reason, stem plots are most useful for fairly small data sets. Study the stem plot on page 59 of your book. Make sure you understand how to read the values.

The example on page 60 of your book shows how to use the data in a stem plot to create a histogram. The values in the stem plot are years. The stem values show the first three digits of the year, and the leaves show the last digit. So, for example, the values next to the stem 184 represent the years 1840, 1841, 1843, and 1848.

Read through the example to see how the histogram was created. Then, enter the data into your calculator and experiment with different bin widths. Here are some possibilities.

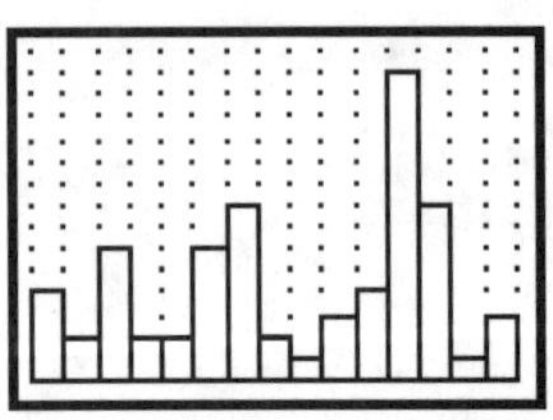

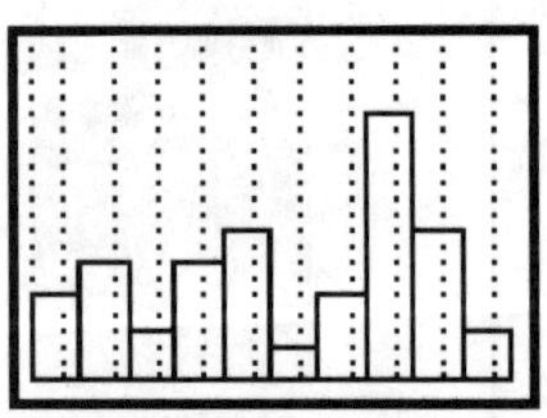

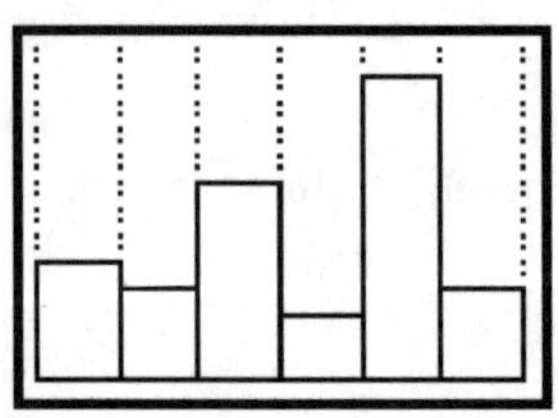

Think about which bin widths best show the spread and distribution of the values.

CONDENSED
LESSON

1.6 Two-Variable Data

In this lesson you will

- practice plotting points on the **coordinate plane**
- create a **scatter plot**
- **look for a relationship** between two variables based on a graph
- use your calculator to create a scatter plot

A **variable** is a quantity—such as pulse rate or mint year—whose value can vary. So far in this chapter, you have been working with **one-variable data.** You have used dot plots, box plots, histograms, and stem plots to display one-variable data.

It is often interesting to look at information about *two* variables to try to discover a relationship. For example, you could collect data about height and shoe size to see if taller people tend to have larger feet than shorter people. Making a **scatter plot** is a good way to look for a relationship in **two-variable data.** Read the information about two-variable graphs on page 67 of your book.

Investigation: Backed into a Corner

This investigation involves collecting and graphing two-variable data.

Steps 1–4 To collect the data, tape two rulers to a desk so that they form an angle. Next, measure the diameter of a circular object. Then, place the object between the rulers so that it fits snuggly, and measure the distance from the vertex of the angle to the point where the object touches the ruler.

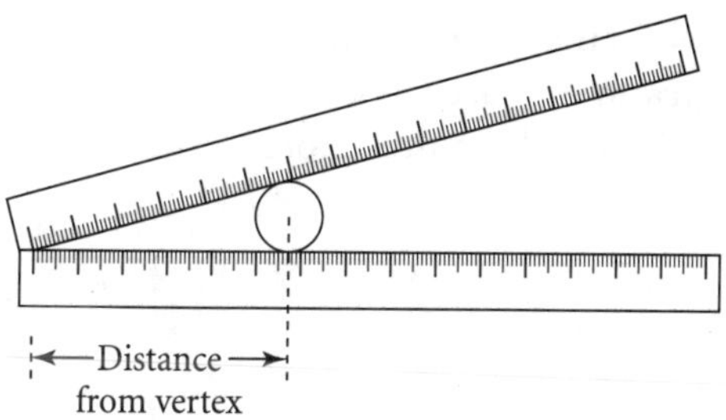

Here are some sample data for this activity.

Diameter (cm)	4.6	6.7	1.9	4.0	7.3	3.1	2.4	5.5
Distance from vertex (cm)	9.6	14.2	3.7	8.8	15.7	6.4	5.4	11.5

Steps 5–7 To see if there might be a relationship between the variables *diameter* and *distance from vertex,* you can make a scatter plot. Before plotting the data, you need to set up and scale the axes to accommodate all the values.

You can put either variable on either axis. In the graph on the following page, the diameter values are on the x-axis and the distance-from-vertex values are on the y-axis. The greatest diameter value is 7.3, and the greatest distance-from-vertex value is 15.7. If you let each grid unit on the x-axis represent 0.5 cm and each grid unit on the y-axis represent 1 cm, all the values will fit on a 16-by-16 grid. Be sure to label the axes with the names of the variables.

(continued)

To plot the data, think of each row as an ordered pair. For example, to graph the first row of values, plot (4.6, 9.6) by moving along the horizontal axis to 4.6 and then straight up until you are even with 9.6. On the right below is the completed scatter plot.

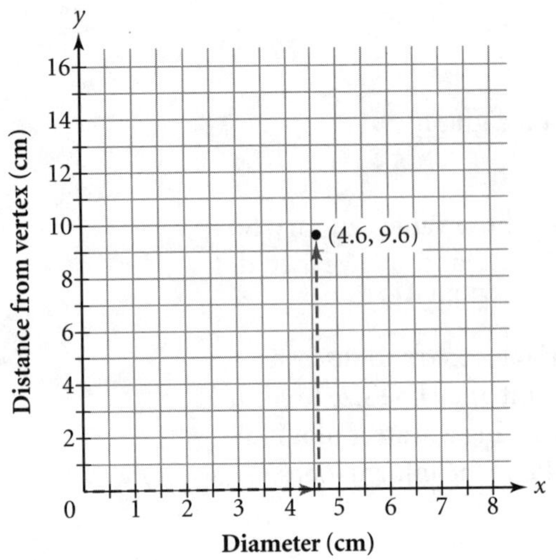

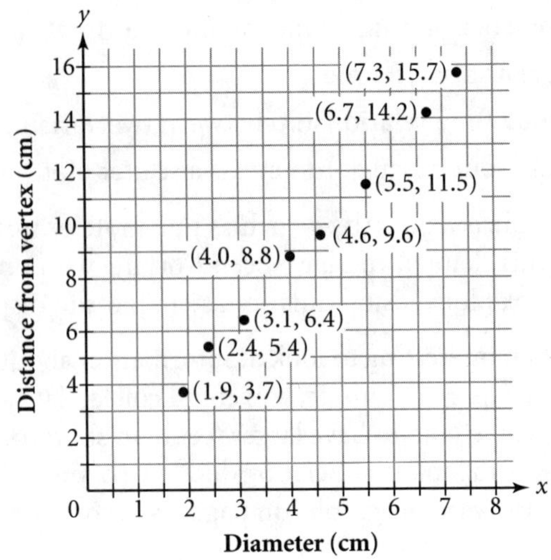

Steps 8–9 Notice that the points form a straight-line pattern. The graph indicates that the greater the diameter of an object, the greater its distance from the vertex will be.

To make a scatter plot on your calculator, you need to enter the data into two lists. Try making a scatter plot of the data above on your calculator. (See **Calculator Note 1E** to learn how to make a scatter plot.)

Review the example on page 69 of your book. The graph in the example, like the graph created in the investigation, is a **first-quadrant graph** because all the values are positive. Sometimes data includes negative values and requires more than one quadrant.

EXAMPLE If you write checks for more money than you have in your account, the account will have a negative balance. The graph at right shows the balance in a checking account at the end of each day over a five-day period. Describe the balance each day.

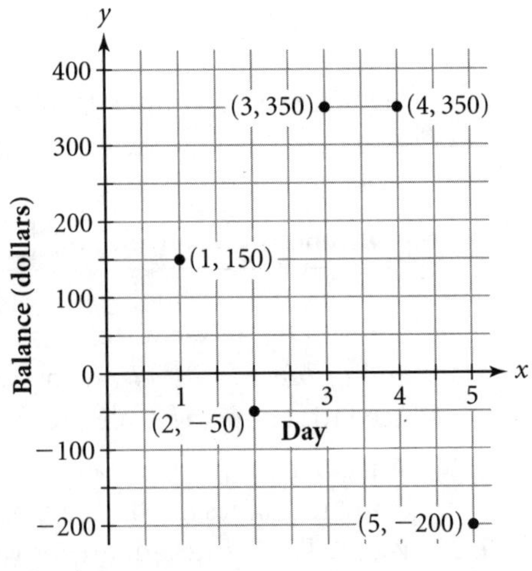

▶ ***Solution*** The point (1, 150) shows that on the first day, the balance was \$150.

The point (2, −50) shows that on the second day, the balance was −\$50.

The points (3, 350) and (4, 350) indicate that on the third and fourth days, the balance was \$350.

The point (5, −200) means that on the fifth day, the balance was −\$200.

CONDENSED LESSON

1.7 Estimating

In this lesson you will

- practice your **estimation** skills
- create a **scatter plot**
- **draw conclusions** about a data set based on graphs
- use your calculator to create a scatter plot

In this lesson you see how to use a scatter plot to compare estimated values to actual values.

Investigation: Guesstimating

Steps 1–3 To test your estimation skills, you can estimate the distances from a starting point to several objects and then measure to find the actual distances. This table shows some sample data. You may want to collect and use your own data, or just add a few of your own values to this table.

Description	Actual distance (m), x	Estimated distance (m), y
Ted's desk	1.86	1.75
Zoe's desk	1.50	1.50
Chalkboard	4.32	3.80
Teacher's desk	1.85	2.00
Blue floor tile	0.57	0.50
Corner of classroom	4.87	4.50
Garbage can	2.35	2.75
Jing's pencil	0.29	0.25
Bookshelf	3.87	4.25

Steps 4–5 You can make a scatter plot to compare the estimated distances with the actual distances. First, set up the axes. Put the actual distances on the x-axis and the estimated distances on the y-axis. Use the *same scale* on both axes.

Plot a point for each pair of values in the table, using the actual distance for the x-value and the estimated distance for the y-value. For example, plot (4.32, 3.80) to represent the data for the chalkboard. Here is the completed graph.

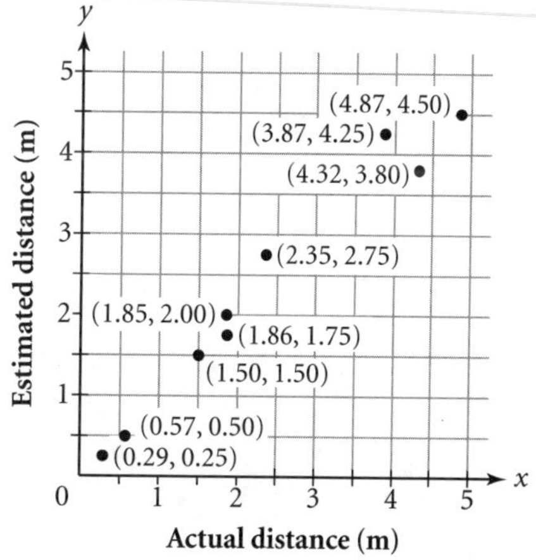

Think about what the graph would look like if *all* of the estimates had been *the same* as the actual measurements. That is, imagine that the graph included (1.86, 1.86), (1.56, 1.56), (4.32, 4.32), and so on. In this case, the points would fall on a straight line.

(continued)

Steps 6–9 Use your graphing calculator to make a scatter plot of the data. Here's what the plot will look like if you use the same scale used in the preceding graph:

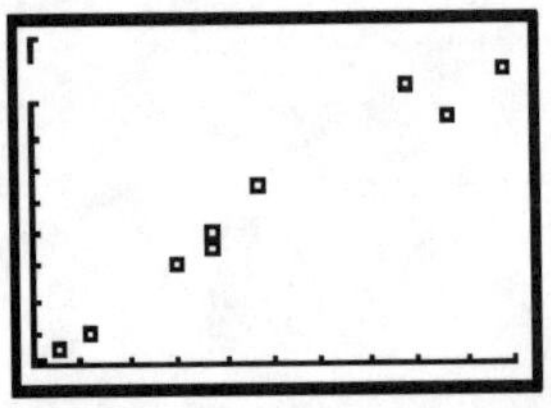

The line $y = x$ represents all the points for which the y-value equals the x-value. In this case, it represents all the points for which the estimated distance equals the actual distance. Graph the line $y = x$ in the same window as your scatter plot. (See **Calculator Note 1H** to graph a scatter plot and an equation simultaneously.)

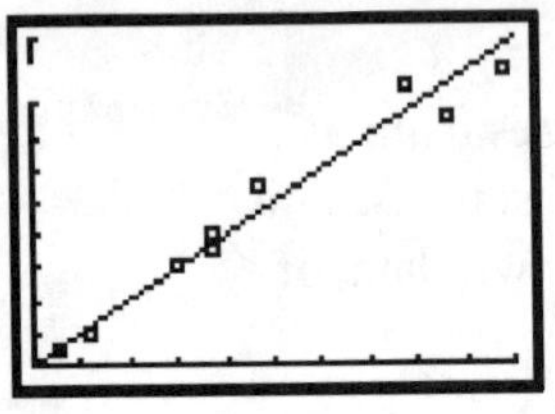

Points that are closer to the line represent better estimates than points that are farther from the line. Use the trace function to look at the coordinates of the points. Notice that points for which the estimate is too low fall below the line and that points for which the estimate is too high fall above the line.

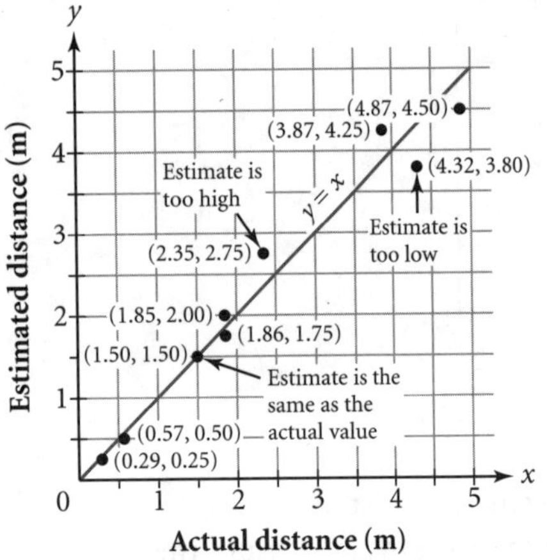

CONDENSED LESSON 1.8

Using Matrices to Organize and Combine Data

In this lesson you will

- use **matrices** to organize and combine data
- use your calculator to **perform matrix calculations**

The table on page 80 of your book shows the average number of hours that people in various countries worked each week in the years 1900 and 1998. Like a table, a **matrix** organizes data into rows and columns. The matrix shows the working-hours data.

$$[A] = \begin{bmatrix} 51.6 & 29.9 \\ 51.7 & 31.7 \\ 51.7 & 38.3 \\ 52.0 & 30.8 \\ 52.0 & 37.9 \\ 52.4 & 35.6 \end{bmatrix}$$

This matrix has **dimensions** 6×2 because it has six rows and two columns.

Work through Example A in your book. In part c, the working-hours matrix is multiplied by 50. Multiplying a matrix by a number means multiplying each entry by that number. So, to find $50 \cdot [A]$, multiply each entry of matrix $[A]$ by 50. Your calculator can do all the multiplications in one step.

To add or subtract two matrices, add or subtract the corresponding entries. Example B shows how to add two matrices. Read through the example carefully and make sure you understand it. In order to add or subtract two matrices, they must have the same dimensions. If the dimensions are different, you won't be able to match up corresponding entries to do the operations.

Now you know how to multiply a matrix by a number and to add or subtract two matrices. As you will see in the investigation, multiplying two matrices is more complicated.

Investigation: Row-by-Column Matrix Multiplication

Steps 1–3 The table and matrix show the costs for large items at two restaurants.

Large-Item Prices

	Pizza Palace	Tony's Pizzeria
Large pizza	\$11.40	\$11.35
Large salad	\$ 3.35	\$ 3.90
Large drink	\$ 2.15	\$ 2.10

$$[D] = \begin{bmatrix} 11.40 & 11.35 \\ 3.35 & 3.90 \\ 2.15 & 2.10 \end{bmatrix}$$

The first column of the matrix gives the prices for Pizza Palace. You can find the total cost for 4 large pizzas, 5 large salads, and 10 large drinks by multiplying the number of each item by the cost for each item.

Pizza Salad Drink Total

$$4 \cdot 11.40 + 5 \cdot 3.35 + 10 \cdot 2.15 = 83.85$$

Number Cost Number Cost Number Cost Cost

(continued)

Steps 4–13 The row matrix $[A]$ and the column matrix $[B]$ contain all the information you need to calculate the total food cost at Pizza Palace. Matrix $[A]$ shows the item quantities, and matrix $[B]$ shows the costs.

$$[A] = [4 \quad 5 \quad 10] \qquad [B] = \begin{bmatrix} 11.40 \\ 3.35 \\ 2.15 \end{bmatrix}$$

Enter $[A]$ and $[B]$ into your calculator and find their *product*, $[A] \cdot [B]$, or

$$[4 \quad 5 \quad 10] \cdot \begin{bmatrix} 11.40 \\ 3.35 \\ 2.15 \end{bmatrix}$$

(See **Calculator Note 1P** to learn how to multiply two matrices.) You should get the 1×1 matrix $[83.85]$. The matrix entry, 83.85, is the total price calculated above. The calculator multiplies each entry in the row matrix by the corresponding entry in the column matrix and adds the results. This is the same calculation.

$$[4 \quad 5 \quad 10] \cdot \begin{bmatrix} 11.40 \\ 3.35 \\ 2.15 \end{bmatrix} = [4 \cdot 11.40 + 5 \cdot 3.35 + 10 \cdot 2.15] = [83.85]$$

You can find the total cost of the food items at Tony's Pizzeria by multiplying the matrices below. Multiply the matrices on your calculator. You should get $[85.90]$.

$$[4 \quad 5 \quad 10] \cdot \begin{bmatrix} 11.35 \\ 3.90 \\ 2.10 \end{bmatrix}$$

You can do one matrix calculation to find the total cost for both restaurants.

$$[4 \quad 5 \quad 10] \cdot \begin{bmatrix} 11.40 & 11.35 \\ 3.35 & 3.90 \\ 2.15 & 2.10 \end{bmatrix}$$

Do this calculation on your calculator. You should get the 2×1 matrix $[83.85 \quad 85.90]$. To get the first entry in the product, the calculator multiplies the row matrix by the first column. To get the second entry, the calculator multiplies the row matrix by the second column. If you try to multiply the matrices in the reverse order, you will get an error message.

To multiply two matrices, the number of columns in the first matrix must be the same as the number of columns in the second because each column entry in the first matrix is multiplied by the corresponding row entry in the second matrix.

Work through Example C in your book to get more practice multiplying matrices.

CONDENSED LESSON

2.1 Proportions

In this lesson you will

- learn several ways for writing a **ratio**
- learn methods for **solving proportions**
- solve problems by writing and solving proportions

The statement "Jackie scored 24 of the team's 64 points" compares two numbers. The **ratio** of the points Jackie scored to the points the team scored is 24 to 64. You can write the ratio as 24 : 64 or as a fraction or decimal. The fraction bar means division, so these expressions are equivalent.

$\frac{24}{64}$ $\quad 24 \div 64 \quad$ 0.375 $\quad \frac{3}{8}$

Read Example A and the text that follows starting on page 93 of your book. Make sure you understand the difference between a **terminating decimal** and a **repeating decimal.**

A **proportion** is an equation stating that two ratios are equal. Here are some true proportions using the integers 3, 5, 9, and 15.

$\frac{9}{15} = \frac{3}{5} \qquad \frac{15}{9} = \frac{5}{3} \qquad \frac{5}{15} = \frac{3}{9} \qquad \frac{15}{5} = \frac{9}{3}$

You can check that the proportions are true by finding the decimal equivalent of each side. The proportion $\frac{3}{15} = \frac{5}{9}$ is not true; 0.2 is not equal to $0.\bar{5}$.

In algebra, a **variable** stands for an unknown number or numbers. In the proportion $\frac{R}{16} = \frac{1}{4}$, you can replace the variable R with any number, but only one number, 4, will make the proportion true.

Investigation: Multiply and Conquer

Steps 1–4 When you multiply both sides of an equation by the *same number,* the sides remain equal. You can use this idea to solve proportions with a variable in one of the numerators. For example, you can solve $\frac{M}{19} = \frac{56}{133}$ by multiplying both sides by 19.

$$\frac{M}{19} = \frac{56}{133}$$

$$19 \cdot \frac{M}{19} = \frac{56}{133} \cdot 19 \qquad \text{Multiply both sides by 19.}$$

$$M = \frac{56}{133} \cdot 19 \qquad \frac{19}{19} \text{ is equivalent to 1.}$$

$$M = 8 \qquad \text{Multiply and divide.}$$

You can check that the solution is correct by replacing M with 8 and making sure the resulting proportion, $\frac{8}{19} = \frac{56}{133}$, is true.

(continued)

Here is the solution to part a in Step 2. Try solving parts b–d on your own.

$$\frac{21}{35} = \frac{Q}{20}$$

$$20 \cdot \frac{21}{35} = \frac{Q}{20} \cdot 20 \qquad \text{Multiply both sides by 20.}$$

$$20 \cdot \frac{21}{35} = Q \qquad \frac{20}{20} \text{ is equivalent to 1.}$$

$$12 = Q \qquad \text{Multiply and divide.}$$

Steps 5–7 In Step 5, the ratios in the proportions from Step 2 have been *inverted.* These new proportions have the same solutions as the original proportions. For example, 12 is a solution to both $\frac{21}{35} = \frac{Q}{20}$ and $\frac{35}{21} = \frac{20}{Q}$. (Check that this is true.) You can use this idea to solve proportions with the variable in a denominator. For example, to solve $\frac{20}{135} = \frac{12}{k}$, just invert the ratios to get $\frac{135}{20} = \frac{k}{12}$, and multiply both sides by 12.

Now, read the question and sample solutions in Step 7 and make sure you understand them.

When a problem involves a ratio or percent, you can sometimes solve it by setting up and solving a proportion. Examples B and C in your book present some sample problems. Here is another example.

EXAMPLE Raj answered 75% of the questions on the algebra midterm correctly. If he got 27 correct answers, how many questions were on the test?

▶ ***Solution*** Let q represent the number of questions on the test. Use the fact that 75% is 75 out of 100 to help you write a proportion. The ratio 27 out of q equals 75 out of 100.

$$\frac{27}{q} = \frac{75}{100} \qquad \text{Write the proportion.}$$

$$\frac{q}{27} = \frac{100}{75} \qquad \text{Invert both sides.}$$

$$27 \cdot \frac{q}{27} = \frac{100}{75} \cdot 27 \qquad \text{Multiply both sides by 27.}$$

$$q = \frac{100}{75} \cdot 27 \qquad \frac{27}{27} \text{ is equivalent to 1.}$$

$$q = 36 \qquad \text{Multiply and divide.}$$

There were 36 questions on the test.

CONDENSED LESSON

2.2 Capture-Recapture

In this lesson you will

- simulate the **capture-recapture method** for estimating animal populations
- write and **solve proportions**
- solve three types of **percent problems:** finding an unknown percent, finding an unknown total, and finding an unknown part

Wildlife biologists use a method called "capture-recapture" to estimate animal populations. This method involves tagging some animals and then releasing them to mingle with the larger population. Later, a sample is taken. Using the ratio of tagged animals in the sample to total animals in the sample, biologists can estimate the animal population.

Investigation: Fish in the Lake

In this investigation a bag of white beans represents a population of fish in a lake. To simulate the capture-recapture method, you reach into the "lake" and remove a handful of "fish." You count the fish in the sample. Instead of putting them back, you replace these fish (white beans) with an equal number of "tagged fish" (red beans).

You then allow the fish to mingle (seal the bag and shake it to mix the beans) and then remove another sample. You count all the fish in the sample and the tagged fish in the sample before you return the fish to the lake. By taking several more samples, you can get a good idea of the ratio of tagged fish to total fish in the lake.

One group of students tagged and returned 84 fish. They then took five samples. Here are their results.

Sample number	Number of tagged fish	Total number of fish	Ratio of tagged fish to total fish
1	8	48	$\frac{8}{48} \approx 0.17$
2	24	102	$\frac{24}{102} \approx 0.24$
3	16	86	$\frac{16}{86} \approx 0.19$
4	17	67	$\frac{17}{67} \approx 0.25$
5	16	75	$\frac{16}{75} \approx 0.21$

To estimate the population of fish in this group's lake, you need to choose one ratio to represent all the samples. You might calculate the median or the mean or use some other method of choosing a representative ratio. For this example, we will use the median of the ratios, which is $\frac{16}{75}$.

(continued)

If the fish were mixed well, the fraction of the tagged fish in a sample should be close to the fraction of the tagged fish in the entire population. In other words, the following should be true:

$$\frac{\textit{tagged fish in sample}}{\textit{total fish in sample}} \approx \frac{\textit{tagged fish in population}}{\textit{total fish in population}}$$

In this case, there were 16 tagged fish in the sample, 75 total fish in the sample, and 84 tagged fish in the population. So you can estimate the fish population, f, by solving this proportion:

$$\frac{16}{75} = \frac{84}{f}$$

To solve the proportion, invert the ratios and multiply both sides by 84.

$$\frac{75}{16} = \frac{f}{84} \qquad \text{Invert both ratios.}$$

$$84 \cdot \frac{75}{16} = \frac{f}{72} \cdot 84 \qquad \text{Multiply both sides by 84.}$$

$$393.75 = f \qquad \text{Multiply and divide.}$$

So there are about 400 fish in the lake (that is, about 400 beans in the bag).

You can describe the results of capture-recapture situations using percents. The examples in your book show three different kinds of percent problems: finding an unknown percent, finding an unknown total, and finding an unknown part. Be sure to read each example and make sure you understand it. Here is one more example.

EXAMPLE In a lake with 350 tagged fish, the recapture results show that 16% of the fish are tagged. About how many fish are in the lake?

► **Solution** In this case, the variable is the total number of fish in the lake, f. Because 16% of the fish are tagged, there are 16 tagged fish to every 100 fish. You can write this as the ratio $\frac{16}{100}$. The ratio of the total number of tagged fish, 350, to the total number of fish in the lake, f, is about $\frac{16}{100}$.

$$\frac{16}{100} = \frac{350}{f} \qquad \text{Write the proportion.}$$

$$\frac{100}{16} = \frac{f}{350} \qquad \text{Invert both ratios.}$$

$$350 \cdot \frac{100}{16} = \frac{f}{350} \cdot 350 \qquad \text{Multiply both sides by 350.}$$

$$2187.5 = f \qquad \text{Multiply and divide.}$$

There are about 2200 fish in the lake.

CONDENSED LESSON

2.3 Proportions and Measurement Systems

In this lesson you will

- find a **conversion factor** to change measurements from centimeters to inches
- use **dimensional analysis** to do conversions involving several steps

If you travel outside the United States, it is helpful to be familiar with the Système Internationale, or SI, known in the United States as the metric system. To change from one system of measurement to another, you can use ratios called **conversion factors.**

Investigation: Converting Centimeters to Inches

To find a ratio you can use to convert centimeters to inches and inches to centimeters, first carefully measure the length or width of several different objects in both units. Here are some sample data. You may want to collect your own data or just add a few measurements to this table.

Object	Measurement in inches	Measurement in centimeters
pen	$5\frac{3}{4} = 5.75$	14.7
calculator	3.0	7.6
paper	$8\frac{1}{2} = 8.5$	21.6
paper clip	$1\frac{7}{8} = 1.875$	4.7
pencil	$6\frac{13}{16} = 6.81$	17.4
desk	30.0	76.2

Enter the measurements in inches into list L1 of your calculator and the measurements in centimeters into list L2. Enter the ratio of centimeters to inches, $\frac{L2}{L1}$, into list L3, and let your calculator fill in the ratio values. (See **Calculator Note 1I.**) Here is the table for the data above.

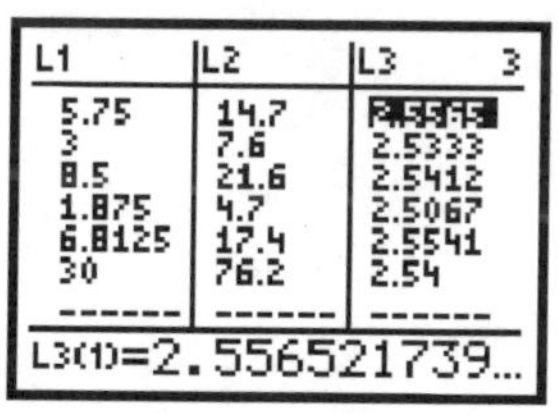

L1	L2	L3
5.75	14.7	2.5565
3	7.6	2.5333
8.5	21.6	2.5412
1.875	4.7	2.5067
6.8125	17.4	2.5541
30	76.2	2.54

L3(1)=2.556521739...

To find a single value to represent the ratio of centimeters to inches, you can use the median or mean of the ratios in list L3. In this case, both the mean and median are about 2.54. So the ratio of centimeters to inches is $\frac{2.54}{1}$ or 2.54. This number is the conversion factor between inches and centimeters. It means that 1 inch is equivalent to about 2.54 centimeters.

(continued)

Using this ratio, you can write and solve a proportion to convert a centimeter measurement to an inch measurement or vice versa. When you set up a proportion, make sure both sides show centimeters to inches or both sides show inches to centimeters.

Here's how to convert 215 centimeters to inches:

$$\frac{2.54}{1} = \frac{215}{x}$$

$$\frac{1}{2.54} = \frac{x}{215}$$

$$215 \cdot \frac{1}{2.54} = \frac{x}{215} \cdot 215$$

$$84.6 \approx x$$

215 centimeters is about 84.6 inches.

Here's how to convert 80 inches to centimeters:

$$\frac{2.54}{1} = \frac{x}{80}$$

$$80 \cdot \frac{2.54}{1} = \frac{x}{80} \cdot 80$$

$$203.2 = x$$

80 inches is about 203.2 centimeters.

Some conversions require several steps. The example in your book shows how to use a strategy called **dimensional analysis** to convert a measurement from feet per second to miles per hour. Here is another example that uses dimensional analysis.

EXAMPLE A car traveled 500 kilometers on 45 liters of gas. Using the facts that 1 mile equals 1.61 kilometers and 1 gallon equals 3.79 liters, express the car's gas consumption in miles per gallon.

▶ ***Solution*** You can use the given information to express the car's gas consumption as the ratio $\frac{500 \text{ kilometers}}{45 \text{ liters}}$. Using the other facts in the problem, you can create fractions with a value of 1, for example, $\frac{1 \text{ mile}}{1.61 \text{ kilometers}}$. By multiplying the original ratio by such fractions, you can convert the gas-consumption ratio to miles per gallon.

$$\frac{500 \text{ \sout{kilometers}}}{45 \text{ \sout{liters}}} \cdot \frac{3.79 \text{ \sout{liters}}}{1 \text{ gallon}} \cdot \frac{1 \text{ mile}}{1.61 \text{ \sout{kilometers}}} = \frac{1895 \text{ miles}}{72.45 \text{ gallons}}$$

$$\approx \frac{26 \text{ miles}}{1 \text{ gallon}} \text{ or 26 miles per gallon}$$

Notice that the fractions equivalent to 1 were chosen so that when the units cancel, the result has miles in the numerator and gallons in the denominator.

CONDENSED LESSON

2.4 Increasing and Decreasing

In this lesson you will

- use proportions to solve problems involving **percent increase** or **percent decrease**
- explore how increasing or decreasing a rectangle's dimensions by a percent affects diagonal length, perimeter, and area

Many everyday situations—including finding sale prices, taxes, and tips—involve working with percent increases or percent decreases. In this lesson you learn how to use proportions to solve such problems.

EXAMPLE To celebrate its 21st anniversary, Elm Street Electronics is offering 21% off all merchandise. Conor wants to buy a pair of speakers originally marked $189. What is the sale price for the speakers?

▸ ***Solution*** If Conor saves 21%, then he pays 100 − 21 or 79% of the original price. Because 79% is the same as 79 out of 100, you can find the sale price, S, by using this proportion.

Sale price ↘ ↙ Part of original price Conor pays

$$\frac{S}{189} = \frac{79}{100}$$

Original price ↗ ↖ The whole original price

To solve the proportion, multiply both sides by 189.

$189 \cdot \frac{S}{189} = \frac{79}{100} \cdot 189$	Multiply both sides by 189.
$S = \frac{79}{100} \cdot 189$	$\frac{189}{189}$ is equivalent to 1.
$S = 149.31$	Multiply and divide.

The sale price is $149.31.

Percent-increase calculations use the same reasoning as percent-decrease calculations. For example, if you add a 15% tip to a dinner bill of $22, you pay 115% of $22.

Example B in your book shows you how to use a calculator to do percent-increase calculations for several prices at the same time. Read the example and make sure you understand it.

Investigation: Enlarging and Reducing

Steps 1–4 At right is a rectangle with one of its diagonals drawn.

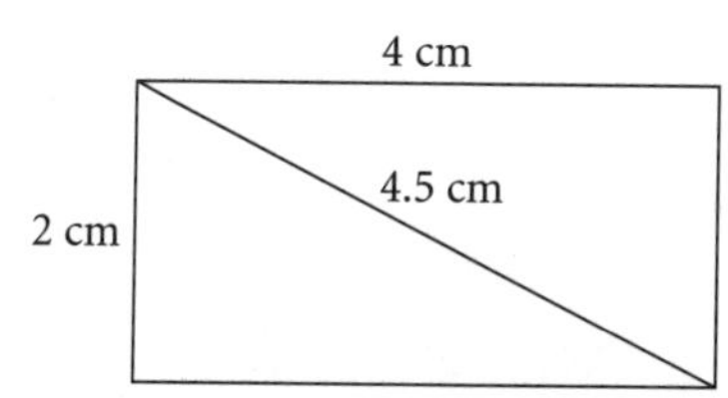

(continued)

If you increase the length and width by 20%, the new dimensions will be 120% of the original dimensions. You can find the length and width of the enlarged rectangle by solving these proportions.

$\frac{l}{4} = \frac{120}{100}$ and $\frac{w}{2} = \frac{120}{100}$

If you decrease the length and width of the original rectangle by 20%, the new dimensions will be 80% of the original dimensions. You can find the length and width of the reduced rectangle by solving these proportions.

$\frac{l}{4} = \frac{80}{100}$ and $\frac{w}{2} = \frac{80}{100}$

Here are the enlarged and reduced rectangles.

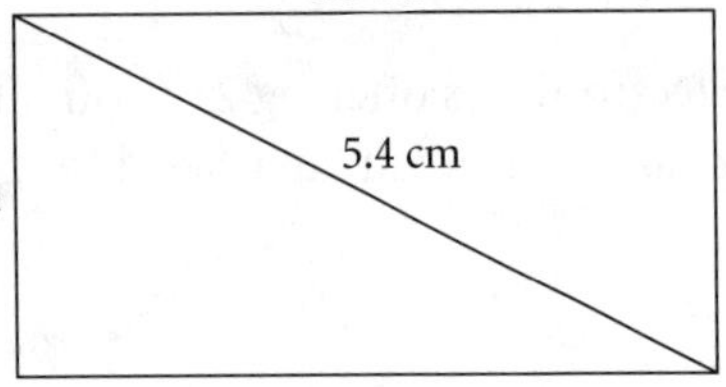

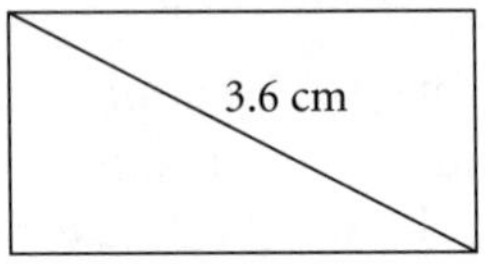

Step 5 This table shows the length, width, diagonal length, perimeter, and area of the rectangles.

The ratio of the diagonal of the enlarged rectangle to the diagonal of the original is $\frac{5.4}{4.5}$ or 1.2 or 120%. If you calculate the ratio of the perimeters, you will also get 1.2 or 120%. So, when the length and width are increased by 20%, the diagonal and the perimeter are increased by 20%.

Rectangle Measurements

	Original dimensions	Enlarged dimensions	Reduced dimensions
Length	4 cm	4.8 cm	3.2 cm
Width	2 cm	2.4 cm	1.6 cm
Diagonal	4.5 cm	5.4 cm	3.6 cm
Perimeter	12 cm	14.4 cm	9.6 cm
Area	8 sq cm	11.52 sq cm	5.12 sq cm

The ratio of the area of the enlarged triangle to the area of the original is $\frac{11.52}{8}$ or 1.44 or 144%. Notice that $1.44 = 1.2^2$. So when the length and width are multiplied by 1.2, the area is multiplied by 1.2^2.

If you calculate similar ratios to compare the reduced rectangle to the original, you will find that when the length and width are decreased by 20% (that is, multiplied by 0.8), so are the diagonal and perimeter. The area is multiplied by 0.8^2 or 0.64.

Step 7 Read the problem posed in Step 7. If the length l and width w of a tabletop are both reduced by half, then the new area A is one-fourth the original area. To see why, do the area calculation for the reduced tabletop.

$A = \frac{1}{2}l \cdot \frac{1}{2}w = \frac{1}{4}lw$

So, if 100 dominoes cover the original tabletop, 25 will cover the reduced tabletop.

CONDENSED LESSON **2.5**

Circle Graphs and Relative Frequency Graphs

In this lesson you will

- create **circle graphs**
- compute **relative frequencies**
- create **relative frequency bar graphs** and **relative frequency circle graphs**

Both bar graphs and circle graphs summarize data that is grouped into categories. **Relative frequency graphs** show the percent of the total value each category represents. In this lesson you will see how to make relative frequency circle graphs and bar graphs.

Investigation: Circle Graphs and Bar Graphs

The bar graph on page 116 of your book shows the land area of each of the seven continents. You can use the graph to approximate the area of each continent and the total area.

Continent	Land area (millions of km²)
Australia	7
Europe	9
Antarctica	14
South America	18
North America	24
Africa	30
Asia	45
Total	147

To convert the data into a circle graph, you need to figure out the angle measure for each section of the graph. To do this, use the fact that there are 360 degrees in a circle.

For example, to find the number of degrees in the section representing Australia, solve this proportion.

Land area of Australia → 7; Angle measure for Australia section → x

$$\frac{7}{147} = \frac{x}{360}$$

Total land area of all continents → 147; Total of angle measures for all sections → 360

The following table shows the angle measure for each section:

Continent	Angle measure
Australia	17°
Europe	22°
Antarctica	34°
South America	44°
North America	59°
Africa	73°
Asia	110°

(continued)

On the left below is the finished graph. To change the graph to a relative frequency circle graph, label each section with the percent of the total area the continent makes up. You can calculate the percents by writing and solving proportions. For example, to find the percent of the total area Australia makes up, you could solve the proportion $\frac{7}{147} = \frac{a}{100}$. On the right is the same circle graph with the sections labeled with percents.

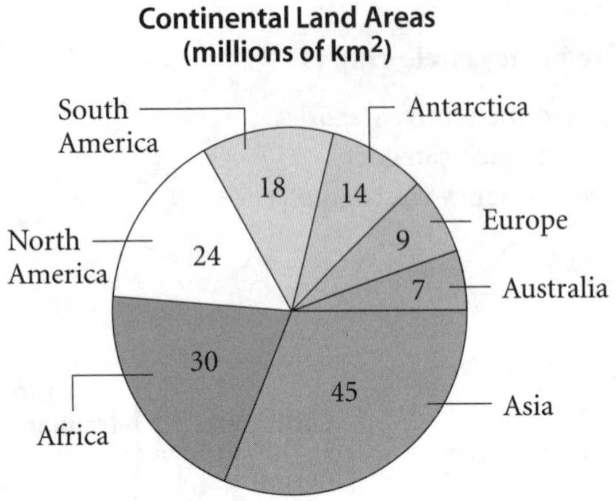

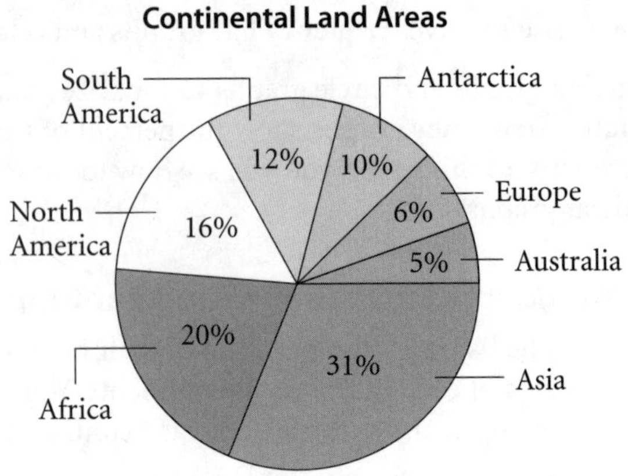

In both circle graphs, the relative size of each section indicates the portion of the total area that continent makes up.

A relative frequency bar graph for this data shows percents of the total land area rather than the land area itself. Here is the completed relative frequency bar graph.

Notice that, like box plots, relative frequency plots do not show actual data values. For example, both of these relative frequency graphs show that Asia makes up 31% of the total land area of the continents, but neither graph indicates what the land area of Asia is.

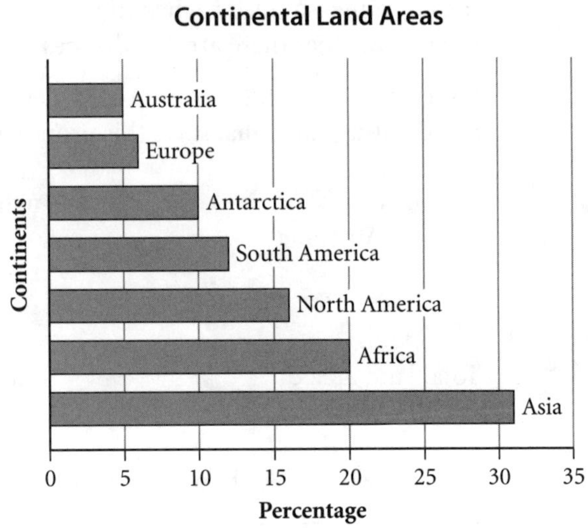

The example in your book walks you through the steps for creating a relative frequency circle graph and bar graph for a different set of data. In the example, a calculator is used to compute the angle measures for the circle graphs and to calculate relative frequencies. Read through this example and follow along on your calculator.

CONDENSED LESSON

2.6 Probability Outcomes and Trials

In this lesson you will

- calculate **observed probabilities** or **relative frequencies** for events
- calculate **theoretical probabilities** for events
- compare relative frequencies to theoretical probabilities

A **probability** of an event is a number between 0 and 1 (or 0% and 100%) that expresses the chance the event will happen. You can find a probability by calculating the ratio of the number of ways the event can occur to the total number of ways under consideration. For example, the probability of tossing a coin and getting heads is $\frac{1}{2}$ because one of the two possible tosses is heads.

EXAMPLE

Joe works in a train station selling coffee and orange juice to commuters. Last Tuesday he sold 60 large coffees, 25 small coffees, 45 large orange juices, and 20 small orange juices. If this distribution accurately reflects the tastes of his customers, what is the probability his first customer next Tuesday will buy a large orange juice? A coffee? Tea?

▶ ***Solution***

The probability that the customer will buy a large orange juice can be expressed as the ratio

$$\frac{\textit{number of customers who bought a large orange juice}}{\textit{total number of customers}} = \frac{45}{150} = 0.3$$

The probability that the customer will buy coffee is

$$\frac{\textit{number of customers who bought coffee}}{\textit{total number of customers}} = \frac{60 + 25}{150} = \frac{85}{150} \approx 0.57$$

None of the customers bought tea because it was not a choice. So the probability that the first customer will buy tea is

$$\frac{\textit{number of customers who bought tea}}{\textit{total number of customers}} = \frac{0}{150} = 0$$

Now, read the example in your book and the text that follows, which explains the terms **trial, outcome, observed probability,** and **theoretical probability.** Think about how each of these terms applies to the example above.

Investigation: Candy Colors

This investigation involves finding the probabilities of selecting various colors from a bag of candy. By conducting an experiment, you can find observed probabilities, or *relative frequencies.* By counting the number of candies of each color, you can find theoretical probabilities.

Steps 1–2 To conduct the experiment, you randomly choose a candy from the bag, record the color, and return the candy to the bag. You repeat this process 40 times. The total number of times each color is selected is called its **experimental frequency.**

(continued)

From the experimental frequencies and the total number of trials (40), you can calculate the observed probability, or relative frequency, of each color. For instance, the observed probability of choosing a red candy will be

$$\frac{\textit{number of red candies drawn}}{\textit{total number of trials}}$$

Here are the data and observed probabilities that one group found.

	Experimental Outcomes						
	Red	**Orange**	**Brown**	**Green**	**Yellow**	**Blue**	**Total**
Tally	III	~~IIII~~ ~~IIII~~ III	~~IIII~~ III	~~IIII~~	~~IIII~~ ~~IIII~~	I	40
Experimental frequency	3	13	8	5	10	1	40
Observed probability (relative frequency)	$\frac{3}{40} = 7.5\%$	$\frac{13}{40} = 32.5\%$	$\frac{8}{40} = 20\%$	$\frac{5}{40} = 12.5\%$	$\frac{10}{40} = 25\%$	$\frac{1}{40} = 2.5\%$	100%

Because the table accounts for every possible color, the probabilities add to 1, or 100%.

Steps 3–6 The group who collected the data above dumped out all the candies and counted the number of each color. They then calculated the theoretical probability of drawing each color. For example,

$$P(\text{red}) = \frac{\textit{number of red candies in the bag}}{\textit{total number of candies in the bag}}$$

Here are their results.

	Outcomes						
	Red	**Orange**	**Brown**	**Green**	**Yellow**	**Blue**	**Total**
Number of candies	2	14	14	10	12	4	56
Theoretical probability	$\frac{2}{56} = 3.6\%$	$\frac{14}{56} = 25\%$	$\frac{14}{56} = 25\%$	$\frac{10}{56} = 17.9\%$	$\frac{12}{56} = 21.4\%$	$\frac{4}{56} = 7.1\%$	100%

In this situation, each candy is equally likely to be selected, but some colors have a higher probability of being chosen than others. The data shows that orange and brown are most likely to be chosen and that red is least likely. Notice that the theoretical probabilities, like the relative frequencies, add to 100%.

Compare the theoretical and experimental probabilities. For example, the observed probabilities predict orange will be chosen about 1 out of 3 times. The theoretical probabilities predict orange will be chosen 1 out of 4 times. In general, the more trials you conduct, the closer the observed probabilities will be to the theoretical probabilities.

CONDENSED LESSON

2.7 Random Outcomes

In this lesson you will

- calculate observed probabilities for a **random** process
- use a calculator to **simulate** coin tosses
- make a graph comparing observed probabilities of tossing heads to the theoretical probability

A process is **random** if you can't predict exactly what will happen on the next trial. Read the introductory text and the example in your book. The example shows that sometimes, even when you cannot predict the exact outcome of a situation, you can use collected results to predict what will happen over the long run. Here is another example.

EXAMPLE One afternoon, Johnna kept track of the number of cars and trucks that passed by her apartment window over an hour period. (She considered minivans and sports-utility vehicles to be trucks.) She counted 72 cars and 40 trucks. Use these results to predict approximately how many of the next 100 vehicles that pass by will be trucks.

▶ ***Solution*** The observed probability that a passing vehicle will be a truck is

$$\frac{\textit{number of trucks}}{\textit{total number of vehicles}} = \frac{40}{112} \approx 36\%$$

Johnna can calculate the *probability* that the next vehicle will be a truck, but she can't know for sure. From Johnna's perspective the event is random. Because the observed probability is 36%, Johnna can expect that 36 of the next 100 vehicles that pass by her window will be trucks.

When you toss a coin once, you cannot predict whether it will show heads or tails because the outcome is random. You do know, however, that there are two equally likely outcomes—heads or tails. Therefore, the theoretical probability of getting a head is $\frac{1}{2}$.

Investigation: Calculator Coin Toss

In this investigation, you use your calculator to **simulate** 100 coin tosses. You then create a scatter plot to compare the theoretical probability of getting heads to an observed probability for 100 trials.

Steps 1–4 Read and follow along with Steps 1–4 in your book. These steps guide you to generate 100 coin tosses. When you are finished, your calculator table will show this information.

- List L1 will show the trial number.
- List L2 will show the result of each toss, with 0 representing tails and 1 representing heads. For example, the table in your book shows that the result of trial 1 was a tail and the result of trial 2 was a head.

(continued)

- List L3 will show the number of heads tossed so far. In the example in your book, three heads were tossed in the first seven trials.
- List L4 will show the observed probability, calculated after each trial. In the example in your book, three heads were tossed in the first seven trials, so the observed probability after seven trials is $\frac{3}{7}$ or about 0.43.

Steps 5–8 You can make a scatter plot showing the observed probability after each trial. Use the values in list L1 (the trial numbers) as the *x*-values and the values in list L4 (the observed probabilities) as the *y*-values. Because there are 100 trials, you will want to set the window to show *x*-values from 0 to 100. Because the highest possible probability is 1, set the maximum *y*-value to 1. Your plot might look something like this.

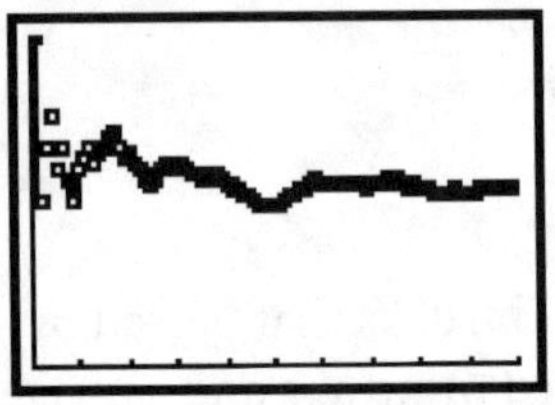

Enter $\frac{1}{2}$, the theoretical probability of tossing a head, into Y1 on the Y= screen. This graphs the line $y = \frac{1}{2}$ on the same screen as your scatter plot.

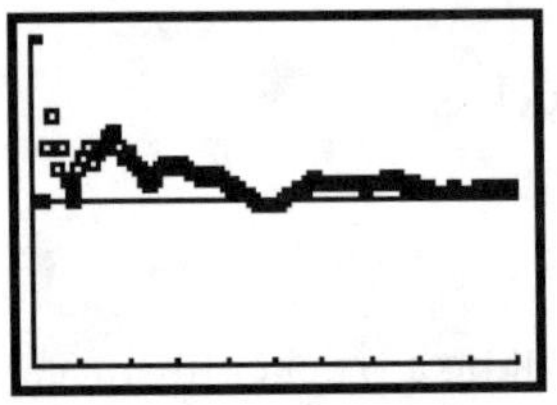

By looking at how close the points are to the line, you can compare the observed probabilities to the theoretical probability. Notice that as the number of trials increases, the points get closer and closer to the line; that is, the observed probabilities get closer and closer to the theoretical probability. Add data for 100 more trials to your table and make another scatter plot. You should observe that the points get even closer to the line.

The more times you toss a coin, the closer the observed probability of getting heads will be to $\frac{1}{2}$. However, even though a pattern emerges in the long run, it will not help you predict a particular outcome. When you flip a coin, you know the theoretical probabilities for heads and tails. In some situations, you cannot calculate theoretical probabilities. In such cases, you can perform many trials and determine observed probabilities based on your experimental results.

CONDENSED LESSON

3.1 Using Rates

In this lesson you will

- learn about a special type of ratio called a **rate**
- use rates to make graphs and tables
- use rates to **make comparisons** and **do calculations**

Investigation: Off to Work We Go

Steps 1–6 Last week, Lacy earned \$300 for 20 hours of work. You can express this as a ratio of pay to hours worked.

$$\frac{\$300}{20\text{ hours}}$$

To find Lacy's pay for 40 hours of work, solve this proportion.

$$\frac{300}{20} = \frac{x}{40}$$

Or, because 40 is $2 \cdot 20$, just multiply \$300, her pay for 20 hours, by 2. Lacy would earn $2 \cdot 300$ or \$600 for 40 hours of work.

To find how much Lacy earns in 1 hour, solve the proportion $\frac{300}{20} = \frac{x}{1}$. Or simply calculate $300 \div 20$. Lacy earns \$15 in 1 hour.

If you write this as $\frac{15}{1}$, you can solve the proportion $\frac{15}{1} = \frac{x}{3}$ to find Lacy's pay for 3 hours of work. Or you can just multiply \$15, her pay for 1 hour, by 3. Lacy earns \$45 in 3 hours.

The table and graph below show Lacy's pay for different numbers of hours. Notice that the points on the graph appear to lie on a straight line.

Lacy's Pay

Time worked (hours)	Pay (\$)
1	15
2	30
3	45
4	60
5	75

Lacy's Pay

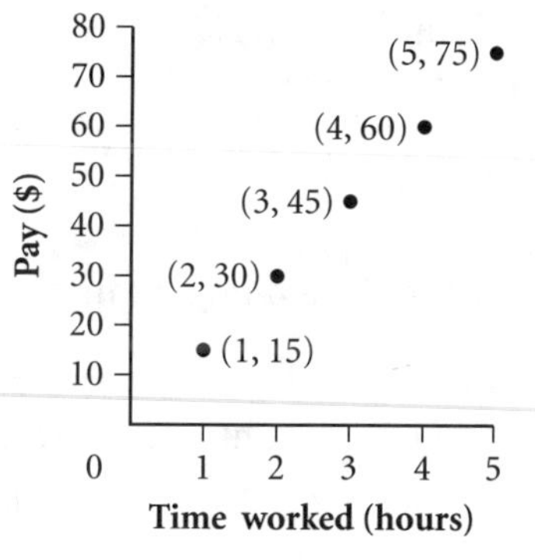

A **rate** is a ratio with 1 in the denominator. Here are three ways to express Lacy's rate of pay.

$\frac{\$15}{1\text{ hour}}$ \$15 for each hour \$15 per hour

(continued)

Steps 7–9 Joseph earned \$513 for 38 hours of work. To find Joseph's rate of pay, write and solve a proportion or find $513 \div 38$. Joseph's rate of pay can be expressed in any of these ways.

$\frac{\$13.50}{1 \text{ hour}}$ \$13.50 for each hour \$13.50 per hour

You can find Joseph's earnings for any number of hours by multiplying the number of hours by the rate.

Create a table like the one for Lacy's pay showing Joseph's pay for from 1 to 5 hours of work. Then, plot your data on the same graph as Lacy's pay. You should notice that Joseph's points also fall along a straight line, but the line is less steep than Lacy's. This is because Lacy's pay rate is higher, so she earns more than Joseph for any number of hours worked.

In the investigation, you discovered that finding rates makes it easy to compare Lacy's pay to Joseph's. Rates also make it easy to do calculations. Rather than solving a proportion, you could just multiply the pay rate by the number of hours worked. Read the rest of the lesson in your book. Then work through the additional example below.

EXAMPLE

On Ali's last phone bill, he was charged \$12.96 for 144 minutes of long-distance calls.

a. How much did Ali pay for each minute?

b. How much would Ali be charged if he talked for 212 minutes?

▶ ***Solution***

a. Express Ali's charges as a ratio: $\frac{\$12.96}{144 \text{ minutes}}$. Divide to find the charge for 1 minute: $12.96 \div 144 = 0.09$. So Ali pays \$0.09 per minute.

b. Multiply the per-minute rate by 212. You can use dimensional analysis to help verify the units for your answer.

$$\frac{\$0.09}{1 \text{ \sout{minute}}} \cdot 212 \text{ \sout{minutes}} = \$19.08$$

So Ali would be charged \$19.08 for 212 minutes.

In the answer to part a, *dollars* is in the numerator and *minutes* is in the denominator. If you switch the quantities in the numerator and the denominator, you get a different rate.

$$\frac{144 \text{ minutes}}{\$12.96} \approx 11 \text{ minutes per dollar}$$

You can use this rate to find the number of minutes Ali could talk for a given number of dollars. For example, for \$2, Ali can talk $11 \cdot 2$ or about 22 minutes.

CONDENSED LESSON

3.2 Direct Variation

In this lesson you will

- represent relationships using graphs, tables, and equations
- use graphs, tables, and equations to find missing data values
- learn about the relationship among **rates, ratios,** and **conversion factors**
- learn about **directly proportional** relationships and **direct variations**

Investigation: Ship Canals

The table on page 146 of your book shows the lengths, in both miles and kilometers, of the world's longest ship canals. Two values are missing from the table. In this investigation you'll learn several ways to find the missing values.

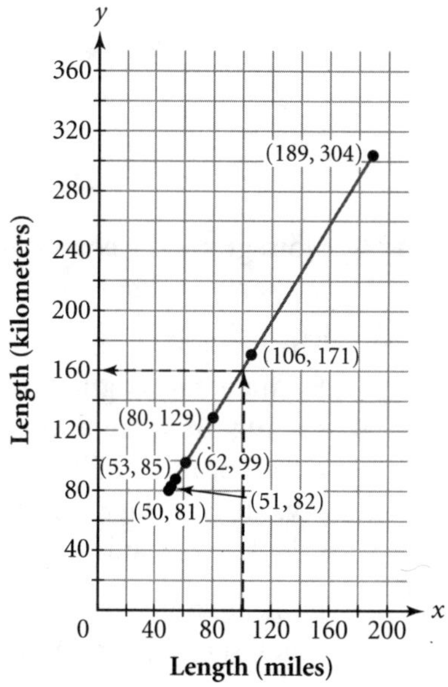

Steps 1–2 This graph shows the data for the first eight canals listed. Connecting the points helps you better see the straight-line pattern.

You can use the graph to estimate the length in kilometers of the Suez Canal. Because the length in miles is 101, start at 101 on the x-axis and move straight up until you reach the line. Then, move straight across to the y-axis, and read off the value. The length is about 160 kilometers. Use a similar method to estimate the length of the Trollhätte Canal in miles.

Steps 3–5 Follow Steps 3 and 4 in your book. When you finish, the List and Graph windows of your calculator should look like this.

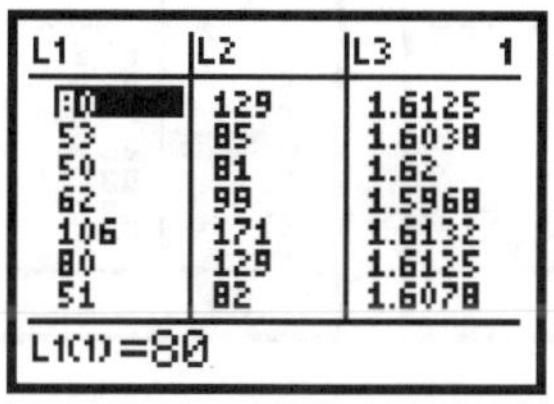

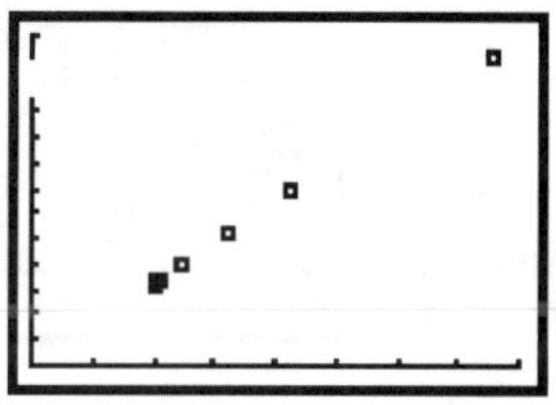

List L3 represents the ratio of kilometers to miles. Each value in this list rounds to 1.6, so there are about 1.6 kilometers in every mile. You can use this conversion factor to find the missing values in the table.

To find the length of the Suez Canal in kilometers, solve this proportion.

$$\frac{1.6 \text{ kilometers}}{1 \text{ mile}} = \frac{t \text{ kilometers}}{101 \text{ miles}}$$

To find the length of the Trollhätte Canal in miles, solve this proportion.

$$\frac{1.6 \text{ kilometers}}{1 \text{ mile}} = \frac{87 \text{ kilometers}}{t \text{ mile}}$$

(continued)

Steps 6–11 There are 1.6 kilometers per mile. So, to change x miles to y kilometers, multiply x by 1.6. You can write this as the equation $y = 1.6x$.

To find the length of the Suez Canal in kilometers, substitute its length in miles for x and solve for y.

$y = 1.6x$ Write the equation.

$y = 1.6 \cdot 101 = 161.6$ Substitute 101 for x and multiply.

To find the length of the Trollhätte Canal in miles, substitute its length in kilometers for y and solve for x.

$y = 1.6x$ Write the equation.

$87 = 1.6x$ Substitute 87 for y.

$\frac{87}{1.6} = x$ To isolate x, divide both sides by 1.6.

$54.375 = x$ Divide.

Now, graph the equation $y = 1.6x$ on your calculator, in the same window as the plotted points.

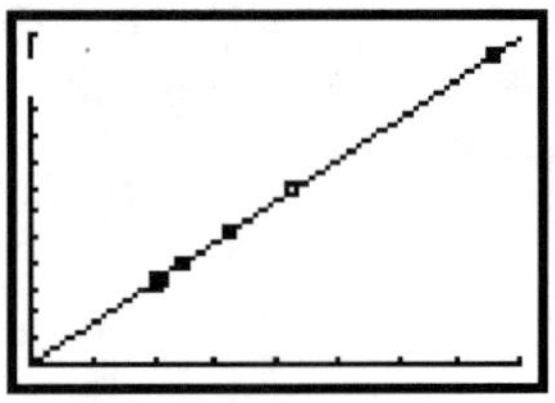

The line goes through the origin because 0 miles = 0 kilometers. Estimate the length of the Suez Canal in miles by tracing the graph and finding the y-value when the x-value is about 101. Then, estimate the length of the Trollhätte Canal in miles by finding the x-value when the y-value is about 87. Your estimates should be close to those you calculated or found using your hand-drawn graph.

Look at your calculator's table display. To find the length of the Suez Canal in miles, scroll down to the x-value 101. The corresponding y-value is 161.6.

To find the length of the Trollhätte Canal in miles, scroll up to show y-values near 87. Using x-increments of 1, the closest y-value to 87 is 86.4. This gives a miles estimate of 54. To find a closer estimate, you can adjust the table setting to show smaller increments.

X	Y1
98	156.8
99	158.4
100	160
101	161.6
102	163.2
103	164.8
104	166.4

X=101

X	Y1
51	81.6
52	83.2
53	84.8
54	86.4
55	88
56	89.6
57	91.2

X=54

You have used several methods for finding the missing values. Which methods do you prefer?

The relationship between kilometers and miles is an example of a type of relationship called a **direct variation.** In a direct variation, the ratio of two variables is constant. Read the text following the investigation carefully. Make sure you understand the terms **directly proportional** and **constant of variation.** Then, read and follow along with the example.

CONDENSED LESSON

3.3 Scale Drawings and Similar Figures

In this lesson you will

- find **scale factors** relating scale drawings to actual objects
- create a **scale drawing**
- use scale factors to find missing lengths in **similar figures**
- write **direct variation** equations relating similar figures

Floor plans and maps are examples of scale drawings. Each has a rate or **scale factor** that relates the measurements of the drawing to the measurements of the real thing the drawing represents.

Investigation: Floor Plans

Steps 1–3 The scale drawing on page 154 of your book shows the floor plan of an apartment. Three of the lengths are labeled with their actual measures. Use a centimeter ruler to measure these labeled lengths. This table compares the actual lengths to the lengths in the drawing.

Actual length (meters)	Length in drawing (centimeters)
6.0	4.0
3.5	2.3
3.3	2.2

Compute the ratio of the actual length to the drawing length for each pair of measurements, and convert each ratio to a decimal. Here is the calculation for the first pair of measurements.

$$\frac{\textit{actual measurement in meters}}{\textit{scale drawing measurement in centimeters}} = \frac{6.0 \text{ meters}}{4.0 \text{ centimeters}}$$
$$= 1.5 \text{ meters per centimeter}$$

Because everything in the floor plan was drawn to the same scale, you should get the same result for each pair of measurements.

Notice that the result above is a *rate*. The rate tells you that there are 1.5 meters in the actual floor for each centimeter in the drawing. So the *scale* for the floor plan is 1 centimeter = 1.5 meters.

Steps 4–5 In the drawing, the bedroom has a length of about 3.7 centimeters and a width of about 2.9 centimeters. (Check this.) To find the actual dimensions in meters, multiply these measurements by 1.5.

$$\text{Actual length} = 3.7 \cancel{\text{ centimeters}} \cdot \frac{1.5 \text{ meters}}{1 \cancel{\text{ centimeter}}} \approx 5.6 \text{ meters}$$

$$\text{Actual width} = 2.9 \cancel{\text{ centimeters}} \cdot \frac{1.5 \text{ meters}}{1 \cancel{\text{ centimeter}}} \approx 4.4 \text{ meters}$$

(continued)

You can use the direct variation equation $y = 1.5x$ to convert any drawing length x in centimeters to the actual length y in meters.

Steps 6–7 The manager of the apartment complex wants to make a model of the apartment in which the 6-meter wall has a length of 10 centimeters. In the scale drawing, this wall has a length of 4 centimeters. You can write this ratio as

$$\frac{\textit{length in model}}{\textit{length in drawing}} = \frac{10 \text{ centimeters}}{4 \text{ centimeters}} = 2.5 \text{ centimeters per centimeter}$$

So to convert measurements from the scale drawing to the scale model, you multiply by 2.5. This can be expressed by the equation $y = 2.5x$, where x is the drawing's length in centimeters and y is the model's length in centimeters.

Now, draw an accurate floor plan for the scale model. In your finished plan, each length should be 2.5 times the floor plan length, and each angle should be the same size as the angle in the floor plan.

The floor plan you drew should have the same *shape* as the original. Figures with the same shape are called **similar figures.** Similar polygons have sides that are proportional and angles that are congruent. Example A in your book shows you how to find the scale factor for a pair of similar figures and then use it to find a missing length. Read through the example and make sure you understand it. Here is another example.

EXAMPLE Zeke is building a model boat. He wants each inch on his model to represent 3.5 feet on the actual boat. Write an equation he could use to convert the actual lengths to model lengths. Explain how Zeke could use a calculator graph to help him make his conversions.

▶ ***Solution*** Let x represent the actual lengths in feet, and let y represent the model lengths in inches. You want to find an equation that lets you calculate y when you know x. You can use the information in the problem to write a proportion and then isolate y to get the equation.

$$\frac{y}{x} = \frac{1}{3.5} \qquad y \text{ corresponds to 1 inch and } x \text{ to 3.5 feet.}$$

$$x \cdot \frac{y}{x} = \frac{1}{3.5} \cdot x \qquad \text{To isolate } y \text{, multiply both sides by } x.$$

$$y = \frac{1}{3.5}x$$

Zeke could enter this equation into his calculator and make a graph. Then he could trace along the line to read each measurement. For example, (14, 4) means that 14 feet on the actual boat should be represented by 4 inches on the model.

CONDENSED LESSON

3.4 Inverse Variation

In this lesson you will

- discover a **fundamental principle of seesaws,** or levers
- study relationships in which two variables are **inversely proportional**
- write equations for **inverse variations**
- use inverse variation equations to solve problems

Investigation: Seesaw Nickels

For two people of different weights to balance on a seesaw, they must sit at different distances from the center. In this investigation you will experiment with a "seesaw" made with a ruler and a pencil. You'll explore different ways of adjusting the weight and distance on one side of a seesaw to balance a fixed weight on the other side.

Steps 1–5 Follow along with Steps 1–4 in your book. When you are finished, your table of data should look something like this.

Left side		Right side	
Number of nickels	Distance from pencil	Number of nickels	Distance from pencil
1	6	2	3
2	3	2	3
3	2	2	3
4	1.5	2	3
6	1	2	3

As you increase the number of nickels on the left side, you must decrease the distance from the pencil in order for the sides to balance. Notice that, for both sides, the product of the number of nickels and the distance is always 6.

Now, stack three nickels 3 inches to the right of the pencil, and repeat Steps 1–4. You will not be able to balance the three nickels with only one nickel. (To do this, you would need to place the nickel 9 inches from the pencil, which isn't possible.) However, in all other cases, you should find that the product of the number of nickels and the distance for the left side is always 9, the same as the product on the right side.

In general, the product of the left nickels and the left distance is equal to the product of the right nickels and the right distance. If you let N and D represent the left nickels and left distance and n and d represent the right nickels and right distance, you can express this as an equation.

$$N \cdot D = n \cdot d$$

The text at the top of page 165 in your book expresses the relationship you discovered as a multiplication equation and as two proportions. You can use what

(continued)

you know about solving proportions to show that the three equations are equivalent. For example, to show that the first proportion is equivalent to the multiplication equation, multiply both sides of the proportion first by *right nickels* and then by *left distance.* (Try it!)

In the seesaw investigation, the product of the left distance and left nickels was constant. Read Example A in your book carefully. It discusses another relationship in which two variables have a constant product. Such a relationship is called an **inverse variation,** and the variables are said to be *inversely proportional.*

The equation for an inverse variation can be written in the form $xy = k$ or $y = \frac{k}{x}$, where x and y are the inversely proportional variables and k is the constant product, called the **constant of variation.** The graph of an inverse variation is always curved and never crosses the x- or y-axis.

EXAMPLE

The time (in hours) it takes to travel a fixed distance (in miles) is inversely proportional to average travel speed (in miles per hour). When Tom rides his bike at an average speed of 12 miles per hour, it takes him 0.25 hour to get from home to school.

a. How long will it take him to get from home to school if he rides his scooter at an average speed of 8 miles per hour?

b. If he has 45 minutes to get to school, what's the slowest he could travel and still get to school on time?

▸ Solution

a. Let t be Tom's travel time, and let r be his travel speed. Because t and r are inversely proportional, they have a constant product k. Because $r = 12$ when $t = 0.25$, k must be $12 \cdot 0.25$ or 3. You can now write the inverse variation equation as $t = \frac{3}{r}$.

Use the equation to answer the question.

$t = \frac{3}{8}$ Substitute 8 for r.

$t = 0.375$ Divide.

So it would take Tom 0.375 hour, or 22.5 minutes.

b.

$0.75 = \frac{3}{r}$ t is 0.75 hour.

$\frac{1}{0.75} = \frac{r}{3}$ To get r in the numerator, invert both ratios.

$3 \cdot \frac{1}{0.75} = \frac{r}{3} \cdot 3$ To isolate r, multiply both sides by 3.

$4 = r$ Multiply and divide.

The slowest Tom could travel is 4 miles per hour.

CONDENSED LESSON

4.1 Order of Operations and the Distributive Property

In this lesson you will

- apply the **order of operations** to evaluate expressions
- use the **distributive property** to do mental math
- use the distributive property to rewrite **algebraic expressions**

The order of operations specifies the order in which the operations in an expression should be evaluated. For example, to evaluate $12 - 2 \cdot 5$, you multiply first and then add, so the result is 2. In your text, read the rules for Order of Operations and Example A. Here is another example.

EXAMPLE A Evaluate the expression $7 + \sqrt{13 - 4}$ without a calculator. Then, enter the expression into your calculator to see if you get the same answer.

▶ ***Solution*** The square root symbol is a grouping symbol, so you should evaluate the expression under the square root symbol first.

$$7 + \sqrt{13 - 4} = 7 + \sqrt{9} \quad \text{Subtract the grouped numbers.}$$

$$= 7 + 3 \quad \text{Evaluate the square root.}$$

$$= 10 \quad \text{Add.}$$

For your calculator to recognize the grouping, you need to enter parentheses around the expression under the square root symbol: 7 + [√] (13 – 4) [ENTER].

Investigation: Cross-number Puzzle

In this investigation you practice the rules for order of operations by completing a cross-number puzzle. Read the rules and then complete the puzzle. The finished puzzle is shown here.

You can evaluate the expression $4(8 + 3)$ in two ways.

1. Add first and then multiply: $4(8 + 3) = 4(11) = 44$.
2. Multiply 4 by each number in parentheses and then add the results: $4(8 + 3) = 4 \cdot 8 + 4 \cdot 3 = 32 + 12 = 44$.

This example illustrates the **distributive property.** Read about the distributive property on pages 184 and 185 of your book, and then carefully work through Examples B–D. Here are two additional examples.

(continued)

EXAMPLE B Find $8 \cdot 97$ without using your calculator.

▶ ***Solution*** Here are two ways to find the product.

Think of 97 as $90 + 7$. **Find $8 \cdot 90$ and add $8 \cdot 7$.**	**Think of 97 as $100 - 3$.** **Find $8 \cdot 100$ and subtract $8 \cdot 3$.**
$8 \cdot 97 = 8(90 + 7)$	$8 \cdot 97 = 8(100 - 3)$
$= 8 \cdot 90 + 8 \cdot 7$	$= 8 \cdot 100 - 8 \cdot 3$
$= 720 + 56$	$= 800 - 24$
$= 776$	$= 776$

EXAMPLE C The art museum offers students a \$3 discount on the admission price.

a. Let A represent the regular admission price. Write two equivalent expressions for the total price for the 25 students in Ms. Dahl's class to go to the museum. One expression should use parentheses and the other should not.

b. Use your calculator to check that both expressions you wrote in part a give the same total admission price.

▶ ***Solution***

a. One expression for the total price is $25(A - 3)$. You can rewrite the expression without parentheses by distributing the 25 to get $25A - 75$.

b. Enter this list of possible admission prices in list L1: {5, 8, 10, 12}. The calculator screen shows the list of prices stored in list L1. Then it shows the two expressions (using list L1 in place of the variable A) evaluated. Both expressions give the same results. The cost of taking the class to the museum is \$50 if the admission price is \$5, \$125 if the admission price is \$8, and so on.

```
{5,8,10,12}→L1
        {5 8 10 12}
25(L1-3)
{50 125 175 225}
25L1-75
{50 125 175 225}
```


CONDENSED LESSON **4.2**

Writing Expressions and Undoing Operations

In this lesson you will

- use **algebraic expressions** to explain number tricks
- work backward to **solve equations**

Read about the number trick described in the first paragraph of the lesson in your book. Try the trick with several different starting numbers. You should always get 7. You can use algebra to figure out how this trick works.

Investigation: The Math Behind the Number Trick

Steps 1–3 Follow the instructions in Step 1 of your book to try the trick on several numbers at the same time. Notice that the last calculator operation, Ans – L_1, is different from the others because it involves the number you started with, rather than a constant value.

Number tricks like this work because certain operations get "undone" in the course of the trick. In this trick, the step Ans/3 undoes Ans · 3.

Page 191 of your book shows a scheme for representing a different number trick. The symbol [+1] represents one positive unit, and the symbol [*n*] represents a variable. You can think of [*n*] as a container for different unknown starting numbers. The right column below describes each stage of the trick in words.

Stage		Description
1	(*n*)	Pick a number.
2	(*n*) [+1] [+1] [+1]	Add 3.
3	(*n*) (*n*) [+1] [+1] [+1] [+1] [+1] [+1]	Multiply by 2.
4	(*n*) (*n*) [+1] [+1]	Subtract 4.
5	(*n*) [+1]	Divide by 2.
6	[+1]	Subtract the original number.
7	[+1] [+1] [+1]	Add 2 (or multiply by 3).

Steps 4–6 Beginning at Stage 6, everyone who tries the trick will get the same result because the number they picked is subtracted in Stage 5. Use your calculator to test this trick with a list of starting numbers. You should get a result of 3 no matter what number you start with.

Step 7 Invent your own number trick. You might find it helpful to draw diagrams like those above. The diagram for the final stage should not contain any variables.

Work through Example A in your book. It shows how to write an algebraic expression to represent the stages of a number trick. The final expression summarizes all the stages of the trick in symbolic form.

(continued)

Here is an expression for the number trick on page 190.

$$\frac{3(x+9)-6}{3} - x$$

Make sure you understand how the expression matches the trick. Notice that the division by 3 undoes the multiplication by 3 and that the x you start with is subtracted at the end.

Now, read Example B in your book. It shows how to write the number trick described by an algebraic expression.

An **equation** is a statement that says the value of one number or algebraic expression is equal to the value of another number or algebraic expression. For example, this equation represents the number trick at the beginning of the lesson.

$$\frac{3(x+9)-6}{3} - x = 7$$

The value of a variable that makes an equation true is a **solution** to the equation. For the equation above, *every* number is a solution. However, this is not usually the case. For example, 4 is the only solution to $2x + 3 = 11$.

Work through Example C in your book. Part c shows how to solve an equation by working backward, undoing each operation until you reach the solution. Here is another example.

EXAMPLE Solve $\frac{2(x+5)}{3} = 6$.

▸ ***Solution*** To get 6, add 5 to a number x, multiply the result by 2, and then divide that result by 3. To find the value of x, start with 6 and work backward, undoing each operation: multiply 6 by 3, divide by 2, subtract 5.

Equation: $\frac{2(x+5)}{3} = 6$

Operations on x	Undo operations	Work backwards
		$x = 4$
$+ (5)$	$- (5)$	9
$\cdot (2)$	$\div (2)$	18
$\div (3)$	$\cdot (3)$	6

The solution is $x = 4$. Check this by substituting it into the original equation.

CONDENSED LESSON

4.3 Recursive Sequences

In this lesson you will

- find **recursive sequences** associated with toothpick patterns
- find missing values in recursive sequences
- write **recursive routines** that generate sequences

A **recursive sequence** is an ordered list of numbers generated by applying a rule to each successive number. For example, the sequence 100, 95, 90, 85, 80, 75, . . . is generated by applying the rule "subtract 5." Example A in your book shows how to use your calculator to generate a recursive sequence. Work through the example and make sure you understand it.

Investigation: Recursive Toothpick Patterns

Steps 1–4 Draw or use toothpicks to build the pattern of triangles on page 200 of your book, using one toothpick for each side of the smallest triangle. For each figure, find the total number of toothpicks and the number of toothpicks in the perimeter.

Build Figures 4–6 of the pattern. This table shows the number of toothpicks and the perimeter of each figure.

	Number of toothpicks	Perimeter
Figure 1	3	3
Figure 2	5	4
Figure 3	7	5
Figure 4	9	6
Figure 5	11	7
Figure 6	13	8

To find the number of toothpicks in a figure, add 2 to the number in the previous figure. To find the perimeter of a figure, add 1 to the perimeter of the previous figure. Below are the recursive routines to generate these number sequences on your calculator.

Number of toothpicks:	Perimeter:
Press 3 [ENTER].	Press 3 [ENTER].
Press +2.	Press +1.
Press [ENTER] to generate each successive term.	Press [ENTER] to generate each successive term.

Build Figure 10, and find the number of toothpicks and the perimeter. Use your calculator routines to check your counts. (The tenth time you press [ENTER], you will see the count for Figure 10.) There are 21 toothpicks in Figure 10 with 12 toothpicks on the perimeter.

(continued)

Steps 5–6 Repeat Steps 1–4 for a pattern of squares. Here is what the pattern should look like.

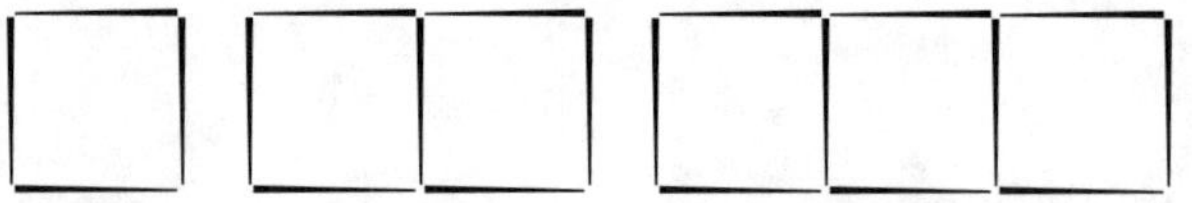

Figure 1 Figure 2 Figure 3

Look for rules for generating sequences for the number of toothpicks and the perimeter of each figure. You should find that the number of toothpicks in each figure is 3 more than the number in the previous figure and that the perimeter of each figure is 2 more than that of the previous perimeter. Notice that if you consider the length of a toothpick to be 1 unit, the area of Figure 1 is 1, the area of Figure 2 is 2, and so on.

Steps 7–8 Create your own pattern from toothpicks and, on your calculator, find recursive routines to produce the number sequences for the number of toothpicks, the perimeter, and the area.

Here is one pattern and the table and recursive routines that go with it.

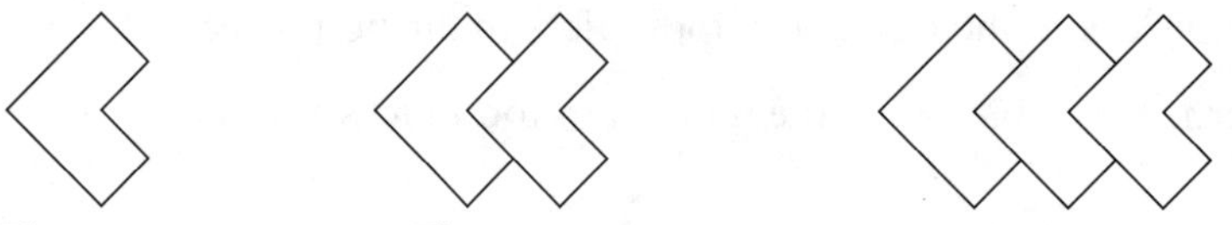

Figure 1 Figure 2 Figure 3

	Number of toothpicks	Perimeter	Area
Figure 1	8	8	3
Figure 2	14	12	6
Figure 3	20	16	9
Figure 4	26	20	12
Figure 12	74	52	36

Here are recursive routines that describe how the figures grow.

Number of toothpicks:	Perimeter:	Area:
Press 8 [ENTER].	Press 8 [ENTER].	Press 3 [ENTER].
Press +6.	Press +4.	Press +3.
Press [ENTER] repeatedly.	Press [ENTER] repeatedly.	Press [ENTER] repeatedly.

For each routine, you can find the result for the figure with 40 puzzle pieces by pressing [ENTER] 40 times. You need 242 toothpicks to build the figure. The perimeter of the figure is 164, and the area is 120.

To find the number of pieces needed for a figure with area 150, use your area routine to generate numbers until you get to 150. You must press [ENTER] 50 times, so you would need 50 pieces. Now, use your number-of-toothpicks routine, pressing [ENTER] 50 times. The result is 302, so you need 302 toothpicks to build the figure.

Now, read Example B in your book, which gives you practice finding missing numbers in recursive sequences.

CONDENSED LESSON

4.4 Linear Plots

In this lesson you will

- use your calculator to apply several recursive routines at once
- graph values generated by recursive routines
- understand how the start value and rule of a recursive routine are reflected in the graph

Follow along with the example on page 206 of your book and make sure you understand it.

Investigation: On the Road Again

Steps 1–3 In your book, read the introduction to the investigation and Step 1. You are given the speed of each vehicle in miles per hour. You can use dimensional analysis to convert each speed to miles per minute (mi/min). For example,

$$\frac{72 \text{ miles}}{1 \cancel{\text{hour}}} \cdot \frac{1 \cancel{\text{hour}}}{60 \text{ minutes}} = \frac{72 \text{ miles}}{60 \text{ minutes}} = 1.2 \text{ miles per minute}$$

Here are the speeds of the three vehicles in miles per minute.

minivan: 1.2 mi/min pickup: 1.1 mi/min

sports car: 0.8 mi/min

Use these speeds to write recursive routines for finding each vehicle's distance from Flint after each minute.

The minivan starts 220 miles from Flint. After each minute, it is 1.2 miles closer to Flint. So the start value is 220, and the rule is "subtract 1.2."

The pickup starts 0 miles from Flint. After each minute, it is 1.1 miles farther from Flint. So the start value is 0, and the rule is "add 1.1."

The sports car starts 35 miles from Flint. After each minute, it is 0.8 mile farther from Flint. So the start value is 35, and the rule is "add 0.8."

To enter the recursive routines on your calculator, enter a list of start values, {220, 0, 35}. Then, apply the rules by entering

{Ans(1) − 1.2, Ans(2) + 1.1, Ans(3) + 0.8}

Use your calculator to find the distance from Flint each minute for the first few minutes. Record your results in a table. Then, change the rules to find distances at 10-minute intervals. To do this, multiply the numbers being added or subtracted by 10. Here are the new rules.

{Ans(1) − 12, Ans(2) + 11, Ans(3) + 8}

(continued)

Time (min)	Minivan (mi)	Sports car (mi)	Pickup (mi)
0	220	35	0
1	218.8	35.8	1.1
2	217.6	36.6	2.2
5	214	39	5.5
10	208	43	11
100	100	115	110

Here is a table with a few values filled in. A complete table would have many more values and show time values until each vehicle reaches its destination.

Steps 4–9 You can plot this information in a graph, with time on the x-axis and distance from Flint on the y-axis. Notice that the points for each vehicle fall on a line. It makes sense to connect the points to represent every possible instant of time.

The start value for each routine is the value where the graph crosses the y-axis. The recursive rule affects how much the distance value changes when the time value increases by 1. This determines the steepness of the line.

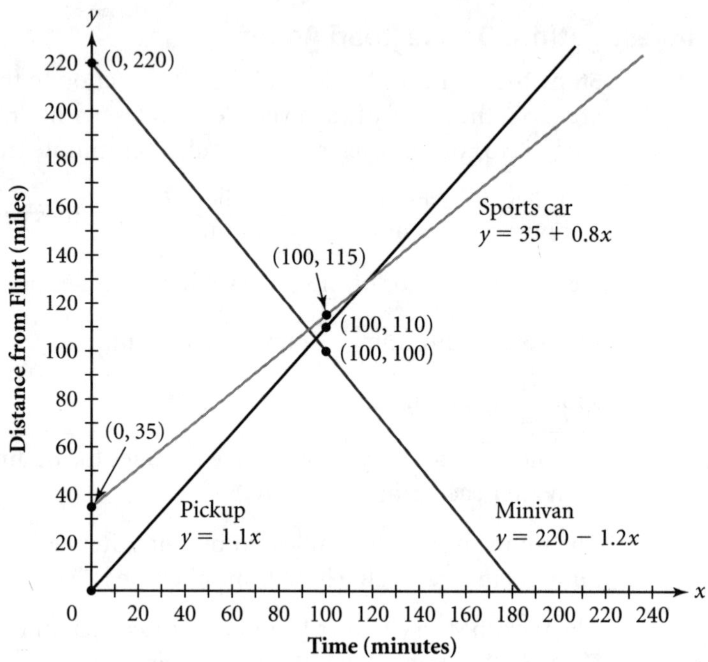

The line for the minivan slants down from left to right because the minivan's distance from Flint is decreasing with time. The lines for the other vehicles slant upward because their distances from Flint are increasing with time.

The lines for the minivan and the sports car cross at about (90, 110). This means that these vehicles pass each other after 90 minutes, when they are both about 110 miles from Flint. At this time, the pickup is about 100 miles from Flint.

The line for the pickup is steeper than the line for the sports car, indicating that the pickup is traveling faster. The lines for the pickup and the sports car cross at about (115, 125), indicating that the pickup passes the sports car after about 115 minutes, when both vehicles are about 125 miles from Flint.

The line for the minivan crosses the x-axis before the lines for the other vehicles reach the 220-mile mark on the y-axis, indicating that the minivan reaches its destination first. The minivan reaches Flint in around 185 minutes. The pickup reaches the bridge in 200 minutes. The sports car reaches the bridge in about 230 minutes.

In this problem, we are assuming the vehicles travel at a constant speed, never stopping or slowing down. Realistically, the vehicles would change speeds, which would be indicated by changes in the steepness of the graph, and they would stop occasionally, which would be indicated by flat portions of the graph. You could not write one recursive routine to generate such graphs; you would have to write different routines for each interval with a different speed.

CONDENSED

LESSON

4.6

Linear Equations and the Intercept Form

In this lesson you will

- **write linear equations** from recursive routines
- learn about the **intercept form** of a linear equation, $y = a + bx$
- observe how the values of a and b in the intercept form relate to the graph of the equation

Investigation: Working Out with Equations

Manisha burned 215 calories on her way to the gym. At the gym, she burns 3.8 calories per minute by riding a stationary bike.

Steps 1–3 You can use the following calculator routine to find the total number of calories Manisha has burned after each minute she pedals.

Press {0, 215} ENTER.

Press Ans + {1, 3.8}.

Press ENTER repeatedly.

In the list {0, 215}, 0 is the starting *minutes* value and 215 is the starting *calories* value. Ans + {1, 3.8} adds 1 to the minute value and 3.8 to the calorie value each time you press ENTER.

You can use your calculator routine to generate this table.

Manisha's Workout

Pedaling time (min), x	Total calories burned, y
0	215
1	218.8
2	222.6
20	291
30	329
45	386
60	443

In 20 minutes, Manisha has burned 215 + 3.8(20) or 291 calories. In 38 minutes, she has burned 215 + 3.8(38) or 359.4 calories. Writing and evaluating expressions like these allows you to find the calories burned for any number of minutes without having to find all the previous values.

Steps 4–7 If x is the time in minutes and y is the number of calories burned, then $y = 215 + 3.8x$. Check that this equation produces the values in the table by substituting each x-value to see if you get the corresponding y-value.

Use your calculator to plot the points from your table. Then, enter the equation $y = 215 + 3.8x$ into the Y= menu and graph it. The line should pass through all the points as shown.

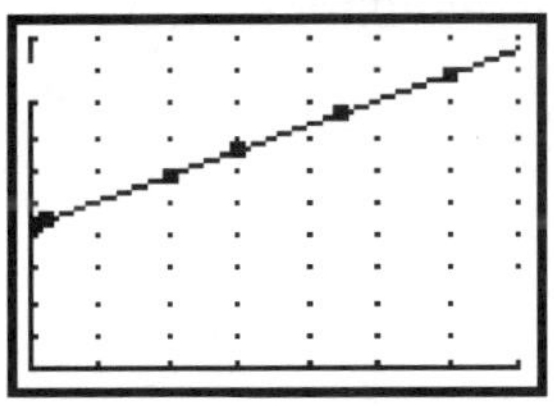

[0, 70, 10, 0, 500, 50]

Note that it makes sense to draw a line through the points because Manisha is burning calories every instant she is pedaling.

(continued)

Steps 8–10 If you substitute 538 for y in the equation, you get $538 = 215 + 3.8x$. You can work backward from 538, undoing each operation, to find the value of x.

$$x \xrightarrow{\times 3.8} 3.8x \xrightarrow{+215} 215 + 3.8x$$

$$\parallel$$

$$85 \xleftarrow{\div 3.8} 323 \xleftarrow{-215} 538$$

Manisha must pedal 85 minutes to burn 538 calories.

Look back at the recursive routine, the equation, and the graph. The starting value of the recursive routine, 215, is the constant value in the equation and the y-value where the graph crosses the y-axis. The recursive rule, "add 3.8," is the number x is multiplied by in the equation. In the graph, this rule affects the steepness of the line—you move up 3.8 units for every 1 unit you move to the right.

In your book, read the text and examples after the investigation. Make sure you understand the **intercept form** of an equation, $y = ax + b$, and how the **y-intercept,** a, and the **coefficient,** b, are reflected in a graph of the equation. Here is an additional example.

EXAMPLE

A plumber charges a fixed fee of $45 for coming to the job, plus $30 for each hour he works.

a. Define variables and write an equation in intercept form to describe the relationship. Explain the real-world meaning of the values of a and b in the equation.

b. Graph your equation. Use your graph to find the number of hours the plumber works for $225.

c. Describe how your equation and graph would be different if the plumber did not charge the $45 fixed fee.

▶ ***Solution***

a. If x represents the hours worked and y represents the total charge, then the equation is $y = 45 + 30x$. The value of a, which is 45, is the fixed fee. The value of b, which is 30, is the hourly rate.

b. Here is the graph. To find the number of hours the plumber works for $225, trace the graph to find the point with y-value 225. The corresponding x-value, 6, is the number of hours.

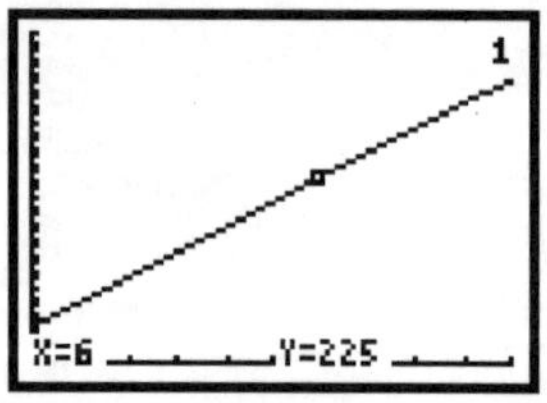

c. If the plumber did not charge a fixed fee, the a-value would be 0 and the equation would be $y = 30x$. The line would have the same steepness, but because the charge for 0 hours would be $0, it would pass through the origin (that is, the y-intercept would be 0).

CONDENSED

LESSON

4.7

Linear Equations and Rate of Change

In this lesson you will

- use the **rate of change** to write a linear equation for a situation
- learn how the rate of change relates to a linear equation and graph
- observe how the *a*-value in $y = a + bx$ relates to the graph

Page 225 of your book shows linear equations in intercept form for some of the situations you have explored in this chapter. For each situation, think about what the variables represent and what the values of a and b mean.

On a cold, windy day, the temperature you feel is colder than the actual temperature due to the wind chill factor. In this lesson you'll look at the relationship between actual temperatures and wind chills. To start, read and follow along with Example A in your book.

Investigation: Wind Chill

Steps 1–4 The table on page 226 of your book relates approximate wind chills to different actual temperatures when the wind speed is 15 miles per hour. Let the input variable, x, be the actual temperature in °F, and let the output variable, y, be the wind chill temperature in °F.

Below is a plot of the data in the viewing window $[-10, 40, 5, -40, 20, 10]$.

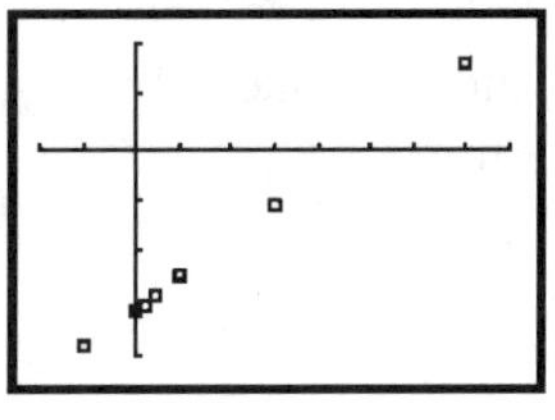

To generate the values on your calculator, you can use the following routine:

Press $\{-5, -38\}$ ENTER.

Press $\{\text{Ans}(1) + 1, \text{Ans}(2) + 1.35\}$.

Press ENTER repeatedly.

The starting list, $\{-5, -38\}$, represents -5°F and its wind chill equivalent. The routine finds wind chill equivalents for temperature of $-5°$, $-4°$, $-3°$, and so on. Each time the actual temperature increases by 1, the wind chill increases by 1.35.

In the table on the next page, columns have been added showing the change in consecutive input and output values and in the rate of change.

(continued)

Input	Output	Change in input values	Change in output values	Rate of change
-5	-38			
0	-31.25	5	6.75	$\frac{6.75}{5} = 1.35$
1	-29.9	1	1.35	$\frac{1.35}{1} = 1.35$
2	-28.55	1	1.35	$\frac{1.35}{1} = 1.35$
5	-24.5	3	4.05	$\frac{4.05}{3} = 1.35$
15	-11	10	13.5	$\frac{13.5}{10} = 1.35$
35	16	20	27	$\frac{27}{20} = 1.35$

Steps 5–8 The rate of change is 1.35, meaning that the wind chill temperature increases by 1.35° for each increase of 1° in the actual temperature. The equation that relates the wind chill x to the actual temperature y is $y = -31.25 + 1.35x$.

The equation $y = -31.25 + 1.35x$ is written in intercept form, $y = a + bx$. Notice that the rule for the recursive routine, "add 1.35," appears as the b-value in the equation. The starting value of the routine, -38, is *not* the value of a in the equation. The value of a is -31.25, the wind chill when the actual temperature is 0°.

Below, the graph of $y = -31.25 + 1.35x$ has been added to the scatter plot. It makes sense to draw a line through the points because every possible temperature has a wind chill equivalent.

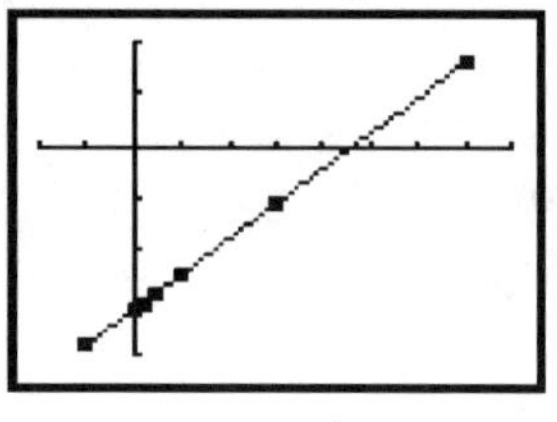

Notice that the y-intercept of the graph, -31.2, is the value of a in the equation.

As you have seen, the rate of change, 1.35, appears as the value of b, or the coefficient of x, in the equation. In the graph, the rate of change is the number of units you move up each time you move 1 unit to the right.

You can use the rate 1.35 to find the actual temperature corresponding to a wind chill of 10.6°. First, note that a wind chill of 16° corresponds to an actual temperature of 35°. To get from a wind chill of 16° to a wind chill of 10.6°, you must subtract 1.35 four times, that is, $10.6 = 16 - 4(1.35)$. Each decrease of 1.35° in wind chill corresponds to a 1° decrease in actual temperature. So the actual temperature corresponding to a wind chill of 10.6° is $35 - 4(1)$, or 31°F.

Example B walks you through a situation similar to the one you looked at in the investigation. Work through this example carefully and make sure you understand it.

CONDENSED LESSON 4.8

Solving Equations Using the Balancing Method

In this lesson you will

- use a balance scale to **model solving an equation**
- solve equations by using the **balancing method**
- compare several methods for solving the same equation

You have found the solutions of linear equations by tracing graphs, by looking at tables, and by working backward to undo the operations. In this lesson you explore how to solve equations using the **balancing method.**

Investigation: Balancing Pennies

The drawing of a balance on page 233 of your book is a visual model of the equation $2x + 3 = 7$. A cup represents the variable x, and pennies represent the numbers. Each cup contains the same number of pennies. To solve the equation, find the number of pennies in each cup.

Steps 1–3 The pictures below show one way to solve the equation. Note that at each stage, the same thing must be done to both sides so that the scale remains balanced.

Picture	Action taken	Equation
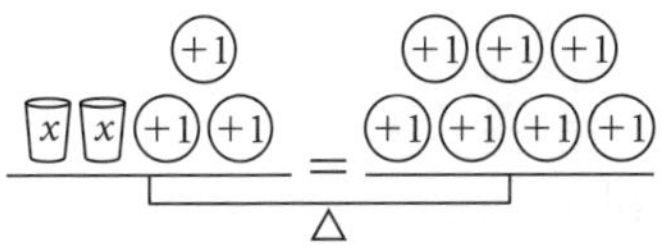	Original balance.	$2x + 3 = 7$
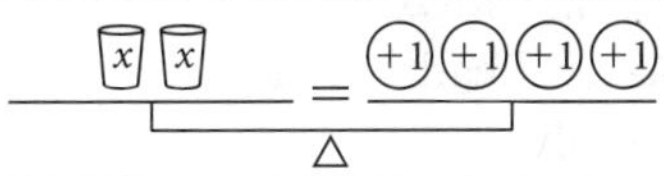	Remove 3 pennies from each side.	$2x = 4$
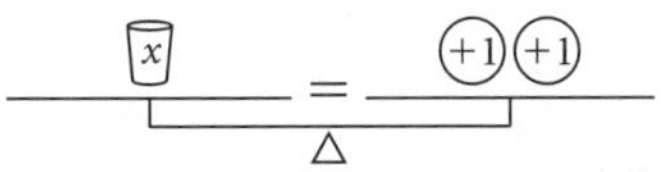	Remove half of each side.	$x = 2$

There are 2 pennies in each cup, so 2 is the solution to the original equation.

Steps 4–8 You can create a pennies-and-cups equation. First, draw a large equal sign and put the same number of pennies on each side. On one side, put some of the pennies into three identical stacks, leaving at least a few pennies, and then hide each stack under a paper cup. Here is the arrangement one group made.

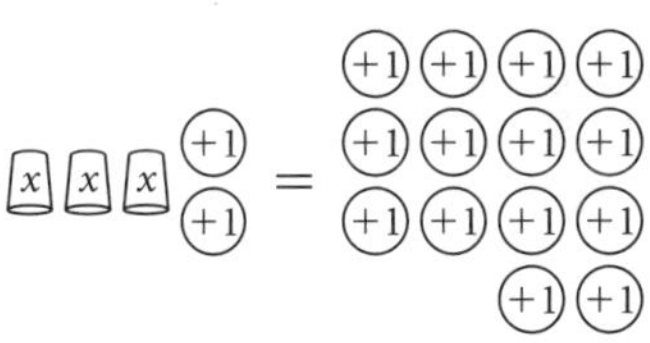

This setup models the equation $3x + 2 = 14$. You can solve the equation—that is, find the number of pennies under each cup—by doing the same thing on both sides of the equal sign. (Think of this as a balance scale; you need to do the same thing to both sides so that the scale remains balanced.)

(continued)

Picture	Action taken	Equation
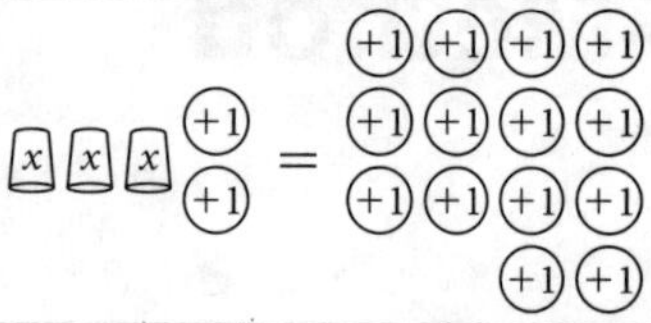	Original setup.	$3x + 2 = 14$
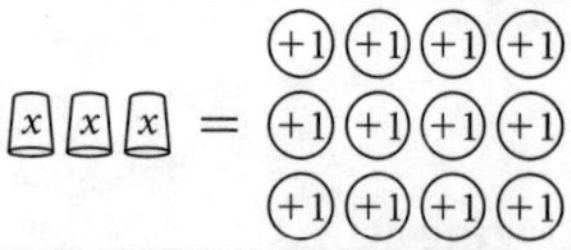	Remove 2 pennies from each side.	$3x = 12$
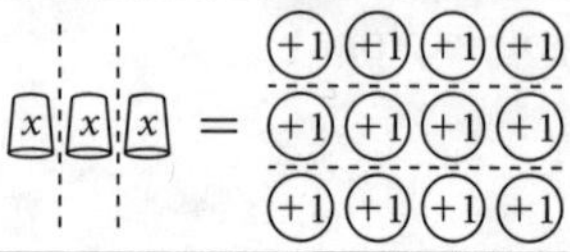	Divide each side by 3.	$\frac{3x}{3} = \frac{12}{3}$
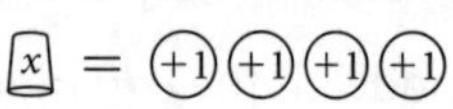	Reduce (leaving a third of each side).	$x = 4$

The solution to $3x + 2 = 14$ is 4. Check this by substituting 4 for x.

The model you used in the investigation works only when the numbers involved are whole numbers. In your book, the text following the investigation and Example A show how you can use a similar model to solve equations involving negative integers. Read this material and make sure you understand it.

Once you get used to doing the same thing to both sides of an equation, you can use the balancing method without drawings or models. This allows you to solve equations involving fractions or negative numbers. Example B in your book shows you how to solve an equation using all four of the methods you know so far. Read through that example. The example below uses the balancing method to solve another equation.

EXAMPLE Solve $7.4 - 20.2x = -1.69$ using the balancing method.

▶ **Solution**

$7.4 - 20.2x = -1.69$	Original equation.
$-7.4 + 7.4 - 20.2x = -1.69 + -7.4$	Add -7.4 to both sides.
$-20.2x = -9.09$	Subtract.
$\frac{-20.2x}{-20.2} = \frac{-9.09}{-20.2}$	Divide both sides by -20.2.
$x = 0.45$	

CONDENSED LESSON

5.1 A Formula for Slope

In this lesson you will

- learn how to calculate the **slope** of a line given two points on the line
- determine whether a point lies on the same line as two given points
- find a point on a line given a known point on the line and the slope

In Chapter 4, you saw that the rate of change of a line can be a numerical and graphical representation of a real-world change like a car's speed. Look at the lines and equation shown on page 251 of your book. Because the coefficient of x represents the rate of change of the line, you can match the equations to the lines by looking at the coefficients—the greater the coefficient, the steeper the line.

The rate of change of a line is often referred to as its **slope.** You can find the slope of a line if you know the coordinates of two points on the line.

Investigation: Points and Slope

Hector pays a flat monthly charge plus an hourly rate for Internet service. The table on page 251 of your book shows Hector's total monthly fee for three months. Because the hourly rate is constant, this is a linear relationship. To find the rate in dollars per hour, divide the change in fee by the change in time for two months. For example, using October and November, you get

$$\frac{28.55 - 19.70}{8 - 5} = \frac{8.85}{3} = 2.95$$

So the rate is \$2.95 per hour. Verify that you get this same result if you use the data for September and October or September and November.

Here is a graph of the Internet data with a line drawn through the points. The arrows show how you can move from (5, 19.70) to (8, 28.55) using one vertical move and one horizontal move.

The length of the vertical arrow is $28.55 - 19.70$ or 8.85 units, which is the change in total fees from October to November. The length of the horizontal arrow is $8 - 5$ or 3 units, which is the change in the number of hours from October to November. Notice that the lengths are the two quantities we divided to find the hourly rate. This hourly rate, 2.95, is the *slope* of the line. The right triangle created by drawing arrows to show vertical and horizontal change is called a **slope triangle.**

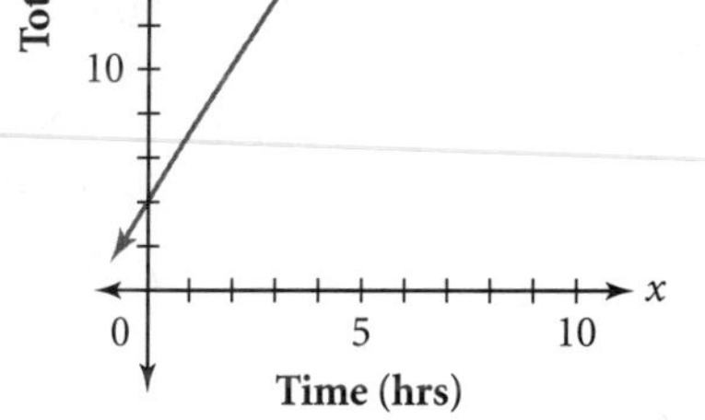

(continued)

At right is the same graph with arrows drawn between the points for September and October. Notice that the slope—or the vertical change divided by the horizontal change—is the same. *Any pair of points on a line will give the same slope.*

You can find the vertical change and the horizontal change by subtracting *corresponding* coordinates. If one of the points is (x_1, y_1) and the other is (x_2, y_2), then the slope is

$$\frac{y_2 - y_1}{x_2 - x_1} \quad \text{or} \quad \frac{y_1 - y_2}{x_1 - x_2}$$

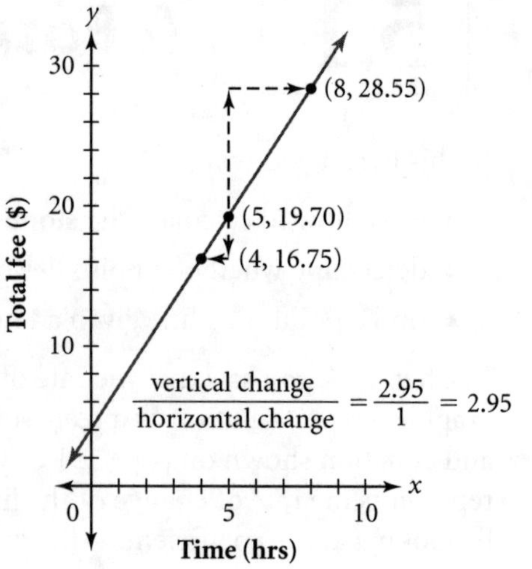

In the investigation the slope of the line is positive. The example in your book involves a line with a negative slope. Read this example carefully. Here is another example.

EXAMPLE Consider the line through $(-2, 3)$ and $(4, -1)$.

a. Find the slope of the line.

b. Without graphing, verify that $(3, -\frac{1}{3})$ is on the line.

c. Find the coordinates of another point on the line.

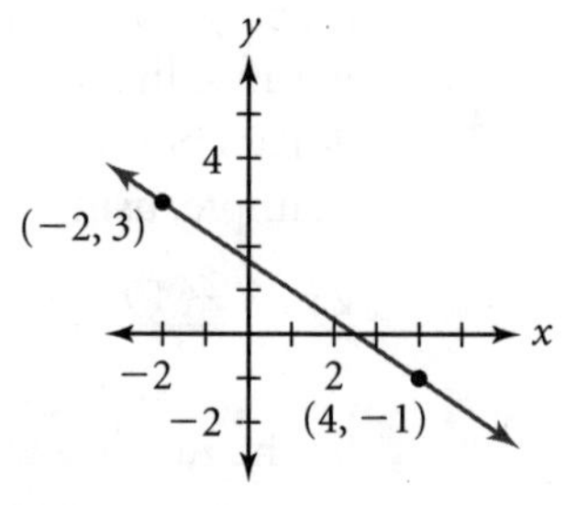

▶ ***Solution***

a. Slope $= \dfrac{-1 - 3}{4 - (-2)} = \dfrac{-4}{6} = -\dfrac{2}{3}$

b. The slope between any two points will be the same. So if the slope between $(3, -\frac{1}{3})$ and either of the original points is $-\frac{2}{3}$, then the point is on the line. The slope between $(3, -\frac{1}{3})$ and $(4, -1)$ is

$$\frac{-1 - \left(-\frac{1}{3}\right)}{4 - 3} = \frac{-\frac{2}{3}}{1} = -\frac{2}{3}$$

So $(3, -\frac{1}{3})$ is on the line.

c. Start with one of the original points. Add the change in x to the x-coordinate and the change in y to the y-coordinate. Let's start with $(-2, 3)$.

$(-2 + \text{change in } x, 3 + \text{change in } y) = (-2 + 6, 3 + (-4)) = (4, -1)$

So $(4, -1)$ is on the line.

Read the rest of Lesson 5.1 in your book. Make sure you understand how you can tell by looking at a line whether the slope is positive, negative, zero, or undefined. Note that when the equation for a line is written in the form $y = a + bx$, the letter b represents the slope.

Writing a Linear Equation to Fit Data

In this lesson you will

- find a **line of fit** for a set of data
- use a **linear model** to make predictions
- learn about the **slope-intercept** form of an equation

Real-world data rarely fall exactly on a line. However, if the data show a linear pattern, you can find a line to model the data. Such a line is called a **line of fit** for the data. Read the information about lines of fit on page 261 of your book.

Investigation: Beam Strength

Steps 1–3 In this investigation, students make "beams" from different numbers of spaghetti strands. Then they test each beam to see how many pennies it can hold before breaking. Here is the data collected by one group of students.

Number of strands	Number of pennies
1	10
2	16
3	28
4	34
5	41
6	46

Steps 4–8 You can make a scatter plot of the data on a graphing calculator or on paper.

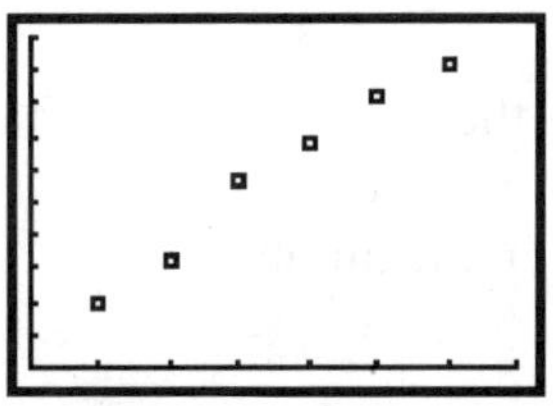

[0, 7, 1, 0, 50, 5]

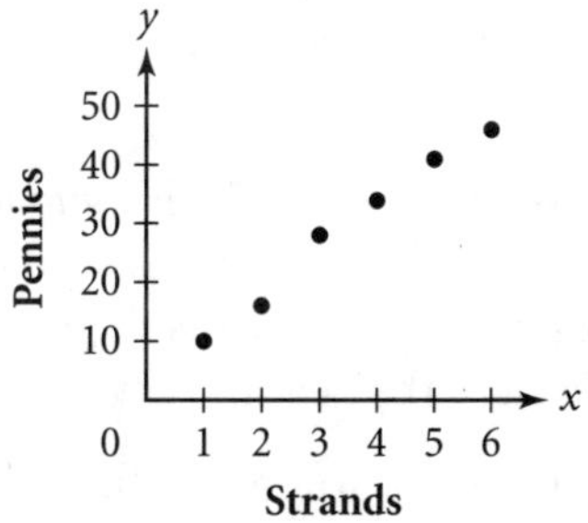

By moving a strand of spaghetti around on your sketch, find a line you think fits the data. Make your line go through two points. Use the coordinates of the two points to find the slope of the line. In the example below, the line goes through (1, 10) and (5, 41). The slope is $\frac{41-10}{5-1}$ or 7.75.

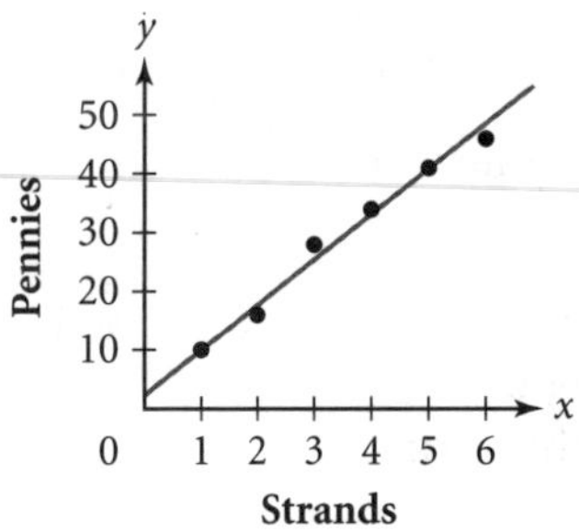

(continued)

Use the slope, b, to graph the equation $y = bx$ on your calculator. For the line above, this is $y = 7.75x$.

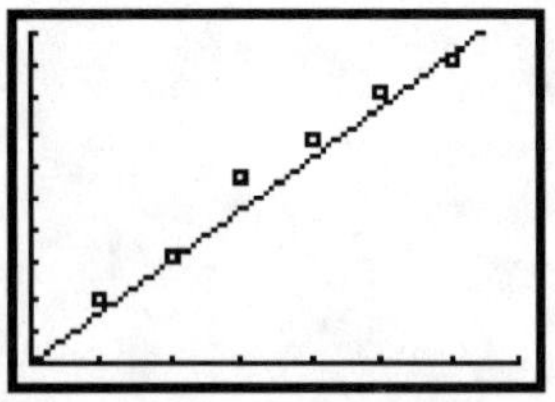

Notice that this line is a little too low to fit the data. Using the sketch above, you can estimate that the y-intercept, a, of the line of fit is about 2. So the equation of the line of fit in intercept form is about $y = 2 + 7.75x$. Graph this equation on your calculator. This line appears to be a good fit. In some cases you'll need to adjust the intercept value several times until you are happy with the fit.

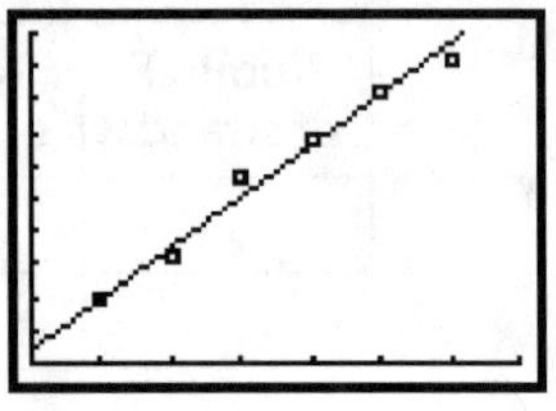

Note that the equation you end up with depends on the two data points your line passes through and your estimate of the y-intercept. Different people will find different equations.

Steps 9–12 The line of fit found above, $y = 2 + 7.75x$, is a *model* for the relationship between the number of strands in the beam, x, and the number of pennies the beam supports, y. The slope, 7.75, represents the increase in the number of pennies each time you add one strand to the beam.

You can use the line of fit to make predictions. To predict the number of strands needed to support \$5 worth of pennies (500 pennies), trace the graph to find the x-value corresponding to a y-value of 500, or solve $500 = 7.75x + 2$. It would take about 64 strands to support this much weight.

The model predicts that a 10-strand beam can hold $7.75(10) + 2$ or about 80 pennies. A 17-strand beam can hold $7.75(17) + 2$ or about 134 pennies.

Now, follow along with the example in your book. It shows you how to fit a line to a different data set.

To find the line of fit in the investigation, we started with the slope and then found the y-intercept. Because of the importance of slope, many people use the **slope-intercept form** of a linear equation, which shows the slope first. In this form, m represents the slope and b represents the y-intercept. The slope-intercept form is $y = mx + b$.

CONDENSED LESSON

5.3 Point-Slope Form of a Linear Equation

In this lesson you will

- write equations in **point-slope form**
- find the equation of a line given one point on the line and the slope
- find the equation of a line given two points on the line

If you are given the slope and the y-intercept of a line, it is easy to write an equation for the line. The example in your book shows you how to find an equation when you know one point and the slope. Here is an additional example.

EXAMPLE When Rosi bought her computer, she made a down payment and then made payments of \$65 per month. After 5 months, she had paid \$450. After 18 months, her computer was paid for. What is the total amount Rosi paid for her computer?

Solution Because the rate of change is constant (\$65 per month), you can model this relationship with a linear equation. Let x represent time in months, and let y represent the amount paid.

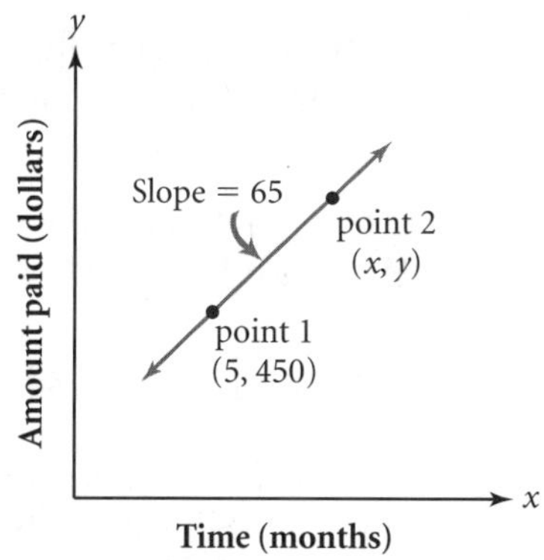

The problem gives the slope, 65, and one point, (5, 450). Let (x, y) represent a second point on the line, and use the slope formula, $\frac{y_2 - y_1}{x_2 - x_1} = b$, to find a linear equation.

$$\frac{y - 450}{x - 5} = 65$$ Substitute the coordinates of the points.

$$(x - 5)\frac{y - 450}{x - 5} = 65(x - 5)$$ Multiply both sides by $(x - 5)$.

$$y - 450 = 65(x - 5)$$ Reduce.

$$y = 450 + 65(x - 5)$$ Add 450 to both sides.

The equation $y = 450 + 65(x - 5)$ gives the total amount paid, y, in x months. To find the total amount Rosi paid in 18 months, substitute 18 for x.

$$y = 450 + 65(x - 5) = 450 + 65(18 - 5) = 450 + 65(13) = 450 + 845 = 1295$$

Rosi paid a total of \$1295 for her computer.

The equation $y = 450 + 65(x - 5)$ is in **point-slope form.** Read about point-slope form on page 271 of your book.

(continued)

Investigation: The Point-Slope Form for Linear Equations

Steps 1–5 Jenny is moving at a constant rate. The table on page 271 of your book shows her distances from a fixed point after 3 seconds and after 6 seconds.

The slope of the line that represents this situation is $\frac{2.8 - 4.6}{6 - 3}$ or -0.6.

If you use the point (3, 4.6), the equation for this situation in point-slope form is $y = 4.6 - 0.6(x - 3)$.

If you use (6, 2.8), the equation is $y = 2.8 - 0.6(x - 6)$.

Enter both equations into your calculator and graph them. You will see only one line, which indicates that the equations are equivalent.

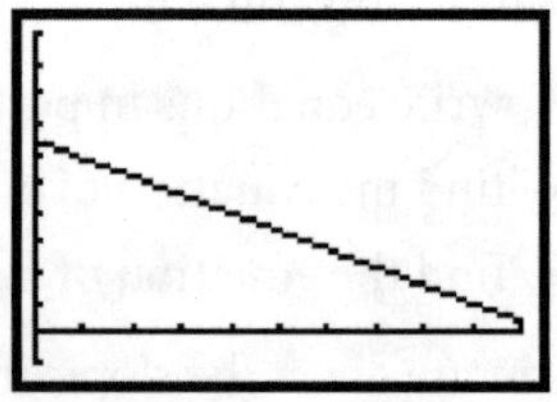

[0, 10, 1, −1, −10, 1]

Now, look at the table for the two equations. Notice that for every x-value, the Y1- and Y2-values are the same. This also indicates that the equations $y = 4.6 - 0.6(x - 3)$ and $y = 2.8 - 0.6(x - 6)$ are equivalent.

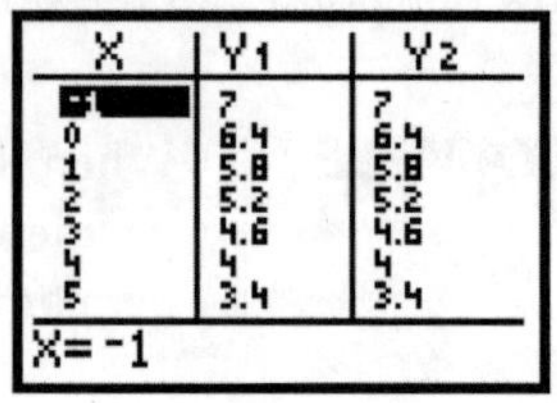

Steps 6–9 The table on page 272 of your book shows how the temperature of a pot of water changed over time as it was heated. If you plot the data on your calculator, you'll see a linear pattern.

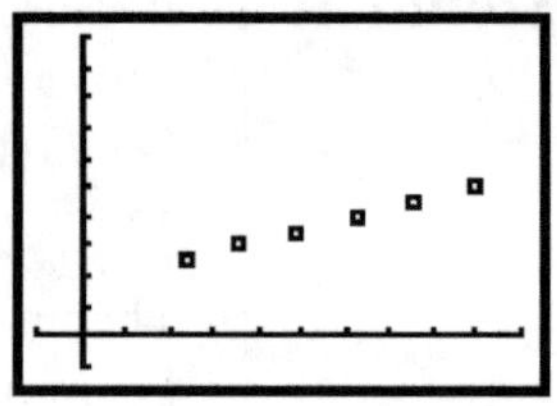

[−10, 100, 10, −10, 100, 10]

Choose a pair of data points. For this example, we'll use (49, 35) and (62, 40). The slope of the line through these points is

$$\frac{40 - 35}{62 - 49} = \frac{5}{13} \approx 0.38$$

Using (49, 35), the equation for the line in point-slope form is $y = 35 + 0.38(x - 49)$. If you graph this equation, you'll see that the line fits the data very closely.

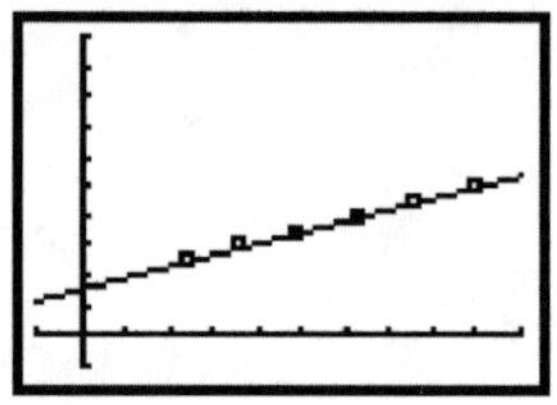

Now, pick a different pair of data points. Find an equation for the line through the points, and graph the equation on your calculator. Does one of the equations fit the data better than the other?

CONDENSED LESSON

5.4 Equivalent Algebraic Equations

In this lesson you will

- determine whether equations are **equivalent**
- rewrite linear equations in **intercept form**
- use **mathematical properties** to rewrite and solve equations

You have seen that the same line can be represented by more than one equation. In this lesson you'll learn how to recognize equivalent equations by using mathematical properties and the rules for order of operations.

Investigation: Equivalent Equations

Page 276 of your book gives six equations. Although these equations look very different, they are all equivalent. To see this, use the distributive property to rewrite each equation in intercept form.

a. $y = 3 - 2(x - 1)$
$= 3 - 2x + 2$
$= 5 - 2x$

b. $y = -5 - 2(x - 5)$
$= -5 - 2x + 10$
$= 5 - 2x$

c. $y = 9 - 2(x + 2)$
$= 9 - 2x - 4$
$= 5 - 2x$

d. $y = 0 - 2(x - 2.5)$
$= 0 - 2x + 5$
$= 5 - 2x$

e. $y = 7 - 2(x + 1)$
$= 7 - 2x - 2$
$= 5 - 2x$

f. $y = -9 - 2(x - 7)$
$= -9 - 2x + 14$
$= 5 - 2x$

All the equations are equivalent to $y = 5 - 2x$. To check that each original equation is equivalent to $y = 5 - 2x$, enter both equations into your calculator and graph them. You will get the same line, which indicates that the same values satisfy both equations.

Now, look at the equations a–o on page 277 of your book. These equations represent only four different lines. To find the equivalent equations, write each equation in intercept form. You should get these results.

Equations a, m, i, and l are equivalent to $y = -5 + 2x$.
Equations b, d, g, and k are equivalent to $y = 2 + 2x$.
Equations e, j, and n are equivalent to $y = -3 - 6x$.
Equations c, f, h, and o are equivalent to $y = 4 - 6x$.

(continued)

In equations h and j, x and y are on the same side of the equation and the other side is a constant. These equations are in **standard form.** Here are the steps for rewriting equation j.

$12x + 2y = -6$	Original equation.
$12x - 12x + 2y = -6 - 12x$	Subtract $12x$ from both sides.
$2y = -6 - 12x$	
$\frac{2y}{2} = \frac{-6 - 12x}{2}$	Divide both sides by 2.
$y = -3 - 6x$	

In the investigation you saw that no matter what form a linear equation is given in, you can rewrite it in intercept form. When equations are in intercept form, it is easy to see whether they are equivalent. Page 278 of your book reviews the properties that allow you to rewrite (and solve) equations. Read these properties and the examples that follow. Here are two more examples.

EXAMPLE C | Is $21x + 3y = 12$ equivalent to $y = -10 - 7(x - 1)$?

▶ ***Solution***

Rewrite $21x + 3y = 12$ in intercept form.

$21x + 3y = 12$	Original equation.
$21x - 21x + 3y = 12 - 21x$	Subtract $21x$ from both sides.
$3y = 12 - 21x$	The expression $21x - 21x$ is 0.
$\frac{3y}{3} = \frac{12 - 21x}{3}$	Divide both sides by 3.
$y = 4 - 7x$	Divide 12 by 3 and $-21x$ by 3.

Rewrite $y = -10 - 7(x - 1)$ in intercept form.

$y = -10 - 7(x - 1)$	Original equation.
$y = -10 - 7x + 7$	Distributive property.
$y = -3 - 7x$	Add -10 and 7.

The equations are not equivalent.

EXAMPLE D | Solve $\frac{4(2x - 3)}{7} = 4$. Identify the property of equality used in each step.

$\frac{4(2x - 3)}{7} = 4$	Original equation.
$4(2x - 3) = 28$	Multiplication property (multiply both sides by 7).
$2x - 3 = 7$	Division property (divide both sides by 4).
$2x = 10$	Addition property (add 3 to both sides).
$x = 5$	Division property (divide both sides by 2).

CONDENSED LESSON

5.5

Writing Point-Slope Equations to Fit Data

In this lesson you will

- write **point-slope equations** to fit data
- use equations of lines of fit to make predictions
- compare two methods for fitting lines to data

This lesson gives you more practice using the point-slope form to model data. You may find that using the point-slope form is more efficient than using the intercept form because you don't have to first write a direct variation equation and then adjust it for the intercept.

Investigation: Life Expectancy

Steps 1–4 The table on page 284 of your book shows the relationship between the number of years a person might be expected to live and the year he or she was born.

Plot the female life-expectancy data in the window [1930, 2010, 10, 55, 85, 5].

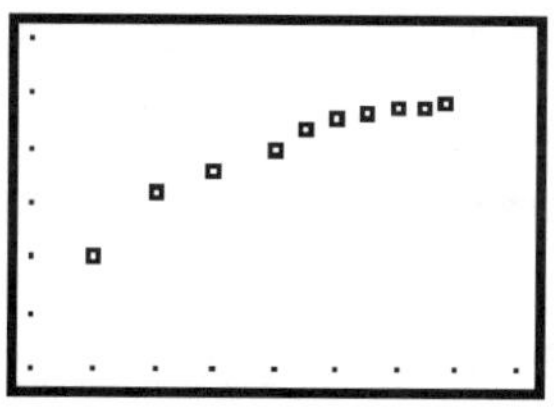

Look for two points on the graph so that the line through the points closely reflects the pattern of all the points. For this example, we'll use (1970, 74.7) and (1990, 78.8). The slope of the line through these points is

$$\frac{78.8 - 74.7}{1990 - 1970} = \frac{4.1}{20} = 0.205$$

Using (1970, 74.7), you can write the equation $y = 74.7 + 0.205(x - 1970)$.

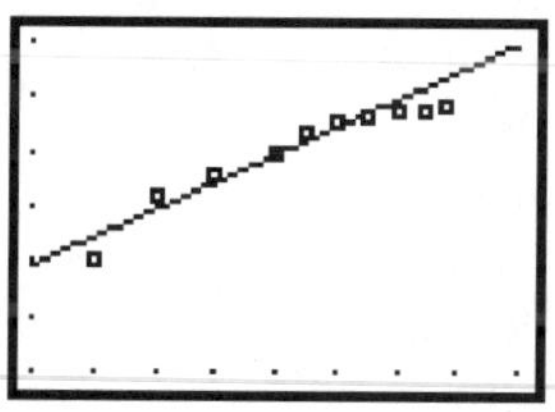

To predict the life expectancy of a female who will be born in 2022, substitute 2022 for x in the equation.

$$\begin{aligned} y &= 74.7 + 0.205(x - 1970) \\ &= 74.7 + 0.205(2022 - 1970) \\ &= 74.7 + 0.205(52) \\ &= 74.7 + 10.66 = 85.36 \end{aligned}$$

The equation predicts that a female born in 2022 will have a life expectancy of 85.36 years.

(continued)

Steps 5–8 If we had chosen a different pair of points, we would have found a different equation and made a different life-expectancy prediction. For example, if we had used (1950, 71.1) and (1995, 78.9), we would have gotten the slope $\frac{78.9 - 71.1}{1995 - 1950} = \frac{7.8}{45} \approx 0.173$ and the equation $y = 71.1 + 0.173(x - 1950)$. This equation gives a life-expectancy prediction of about 83.56 years for a female born in 2022.

Now, find equations of lines of fit for the male data and combined data. Here are equations for all three sets of data, using the points for 1970 and 1990.

Female: $y = 74.7 + 0.205(x - 1970)$

Male: $y = 67.1 + 0.235(x - 1970)$

Combined: $y = 70.8 + 0.23(x - 1970)$

Notice that the slopes of all three lines are close to 0.2, indicating that life expectancy increases by about 0.2 year for each age increase of 1 year, regardless of gender.

The graph below shows all three sets of data and all three lines graphed in the same window. The data for females is plotted with boxes, the data for males is plotted with plus signs, and the combined data is plotted with dots.

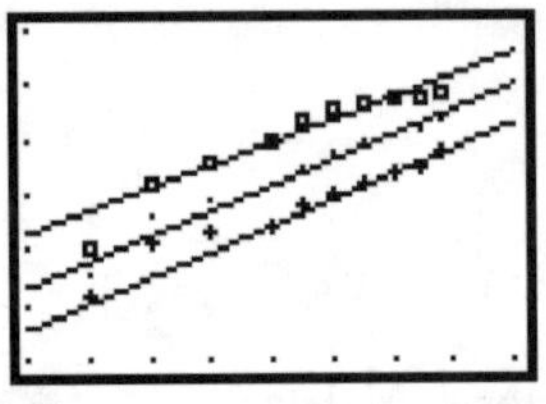

Notice that the line for the combined data is between the other two lines. This is reasonable because the combined life expectancy for both males and females should be between the life expectancy for females and the life expectancy for males.

You have used two methods for finding the equation of a line of fit. One method uses the intercept form, and the other uses the point-slope form. For the intercept-form method (which you used in the "spaghetti beam" investigation), you found a line parallel to the line of fit and then adjusted it up or down, using estimation, to fit the points. With the point-slope method, you get an equation without making any adjustments, but you may find the line does not fit the data as well as you would like.

CONDENSED LESSON

5.6 More on Modeling

In this lesson you will

- use **Q-points** to fit a line to a set of data
- use a line of fit to make predictions

Several times in this chapter you have found the equation of a line of fit for a data set. You probably found that you and your classmates often wrote different equations even though you were working with the same data. In this investigation you will learn a systematic method for finding the equation of a line of fit. This method always gives the same equation for a given set of data.

Investigation: FIRE!!!!

Steps 1–3 In this investigation, students form a line and record the time it takes to pass a bucket from one end of the line to the other. After each trial, some students sit down and the remaining students repeat the experiment. Here are the data one class collected and the graph of the data.

Number of people, x	Passing time (sec), y
22	16
21	18
18	12
16	14
15	12
13	11
12	9
11	7
8	8
5	4

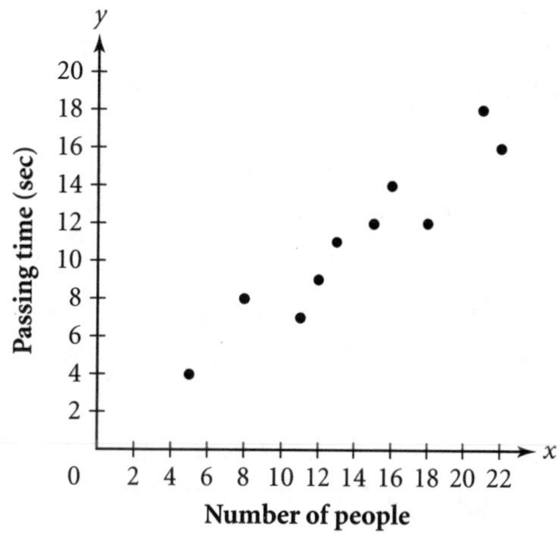

Steps 4–13 You can use the quartiles of the x- and y-values to fit a line to the data. First, find the five-number summaries. The five-number summary for the x-values is 5, 11, 14, 18, 22. The five-number summary for the y-values is 4, 8, 11.5, 14, 18. Use these values to add horizontal and vertical box plots to the graph.

(continued)

Next, draw vertical lines from the Q1- and Q3-values on the x-axis box plot, and draw horizontal lines from the Q1- and Q3-values on the y-axis box plot. These lines form a rectangle. The vertices of this rectangle are called **Q-points.** Q-points may or may not be actual points in the data set. Note that anyone who starts with this data will get the same Q-points.

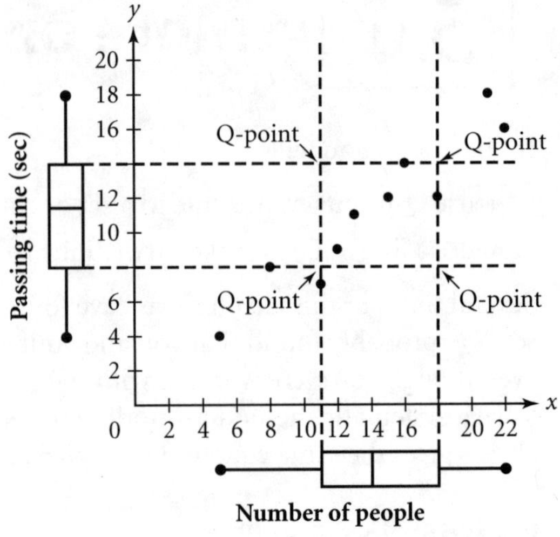

Finally, draw the diagonal of the rectangle that reflects the direction of the data. This is a line of fit for the data.

Use the coordinates of the Q-points $(11, 8)$ and $(18, 14)$ to write an equation for this line. The slope is $\frac{14-8}{18-11}$ or $\frac{6}{7}$, so one equation is $y = 8 + \frac{6}{7}(x - 11)$. In intercept form, this is $y = -1\frac{3}{7} + \frac{6}{7}x$. The slope represents the number of seconds it takes each person to pass the bucket. The y-intercept represents the amount of time it takes 0 people to pass the bucket. In this case, the y-intercept is a negative number, which doesn't make sense in the real situation. This demonstrates that a model only approximates what actually happens.

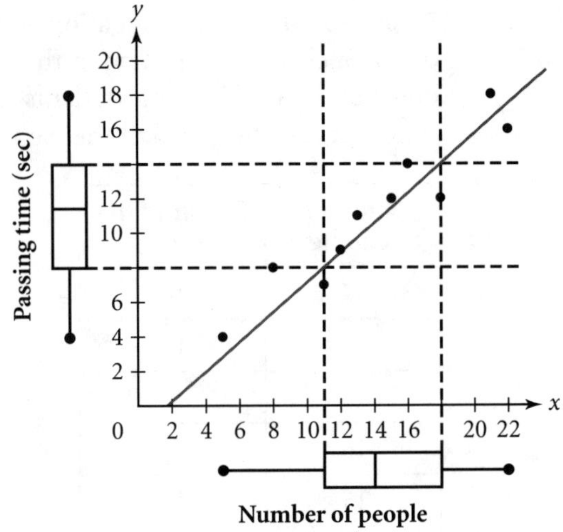

Now, use your calculator to plot the data points, draw vertical and horizontal lines through the quartile values, and find a line of fit. (See **Calculator Note 5B** for help on using the draw menu.)

The example in your book uses the Q-point method to fit a line to a different set of data. Read this example and make sure you understand it.

CONDENSED LESSON

5.7 Applications of Modeling

In this lesson you will

- use and compare three methods of finding a line of fit for a set of data
- use linear models to make predictions

You now know several methods for fitting a line to data. In this lesson you will practice and compare these methods.

Investigation: What's My Line?

The tables on page 296 of your book show how the diameter of a person's pupil changes with age. Table 1 shows daylight pupil diameters. Table 2 shows nighttime pupil diameters. Here, we will work with data in Table 2.

Steps 1–2 First, you'll find a line of fit by "eyeballing" the data. Plot the data on graph paper. Then, lay a strand of spaghetti on the plot so that it crosses the y-axis and follows the direction of the data. Here is an example.

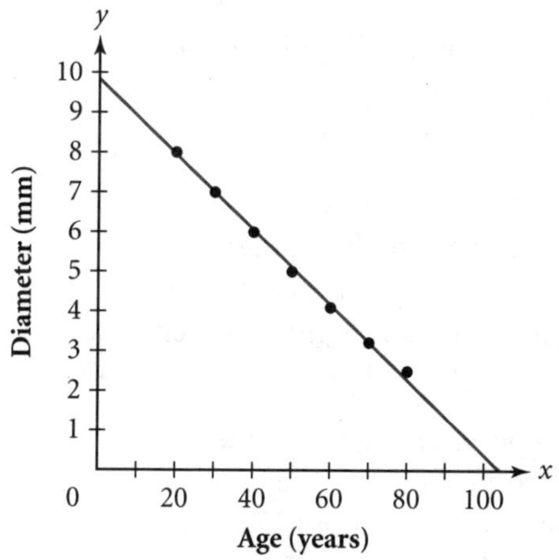

This line crosses the y-axis at about (0, 9.8). The point (30, 7) is also on the line. The slope of the line through these points is $\frac{7-9.8}{30-0}$ or about -0.093. So the equation of this line in intercept form is $y = 9.8 - 0.093x$.

Steps 3–4 Now, you'll fit a line using "representative" data points. Use your calculator with the window [0, 100, 10, 0, 10, 1] to make a scatter plot of the data.

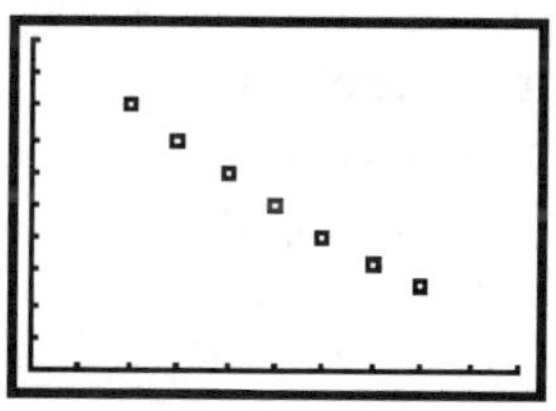

Choose two points you think reflect the direction of the data. For example, you might use (30, 7.0) and (60, 4.1). The slope of the line through these points is $\frac{4.1-7.0}{60-30}$ or about -0.097. An equation for the line through these points in point-slope form is $y = 7 - 0.097(x - 30)$. Rewriting this in intercept form gives $y = 9.91 - 0.097x$.

(continued)

Steps 5–6 Next, you'll find a line of fit using Q-points. The five-number summary of the x-values is 20, 30, 50, 70, 80, so Q1 and Q3 are 30 and 70. The five-number summary of the y-values is 2.5, 3.2, 5.0, 7.0, 8.0, so Q1 and Q3 are 3.2 and 7.0. Use these values to draw a rectangle on the plot. Then, draw the diagonal that reflects the direction of the data. The diagonal passes through (30, 7.0) and (70, 3.2).

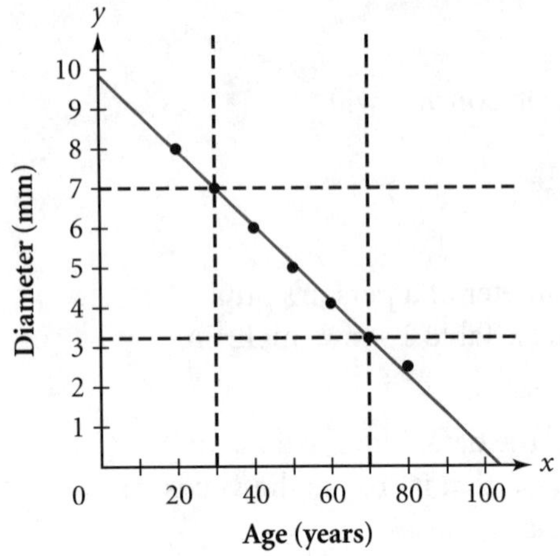

The slope of this line is $\frac{3.2 - 7.0}{70 - 30}$ or -0.095. An equation for this line in point-slope form is $y = 7 - 0.095(x - 30)$. Rewriting this in intercept form gives $y = 9.85 - 0.095x$.

Steps 7–13 In the preceding linear models, the slope indicates the change in pupil diameter for each age increase of 1 year. The three methods gave slightly different values for the slope (-0.093, -0.097, and -0.095). The y-intercept of each model represents the pupil diameter at birth (age 0). These values are also slightly different for the three models (9.8, 9.91, and 9.85.)

You can use any of the models to predict the age when a person's nighttime pupil diameter is about 4.5 mm. For example, to predict a person's age using the first model, $y = 9.8 - 0.093x$, solve $4.5 = 9.8 - 0.093x$. The solution is about 57, so a person's nighttime pupil diameter is 4.5 mm at about age 57.

Changing the y-intercept of an equation increases or decreases every y-coordinate by the same amount. For example, changing the y-intercept of the first model, $y = 9.8 - 0.093x$, to 9.3 gives the equation $y = 9.3 - 0.093x$. The y-coordinate of each point on this second line is 0.5 mm less than the y-coordinate of that point on the original line. For a person of age 45, the first equation gives a pupil diameter of 5.615 mm and the second equation gives a pupil diameter of 5.115 mm.

A small change in slope has little effect on points near the data points, but the effect is magnified for points far out on the line, away from the points. For this data, age values much greater than 80 don't make sense in the real world. However, you might try changing the slope value of one of the models slightly and then substituting age values of 100, 200, and so on to see the effect of the change.

CONDENSED LESSON

6.1 Solving Systems of Equations

In this lesson you will

- represent situations with **systems of equations**
- use tables and graphs to **solve systems of linear equations**

A **system of equations** is a set of two or more equations with the same variables. A solution of a system of equations is a set of values that makes all the equations true. Read the example in your book and then read the example below.

EXAMPLE

The Anytime long-distance plan charges \$4.80 per month plus 5¢ a minute. The TalkMore plan charges 9¢ a minute and no monthly fee. For what number of minutes are the charges for the two plans the same?

a. Write a system of two equations to model this situation.

b. Solve the system by creating a table. Explain the real-world meaning of the solution, and locate the solution on a graph.

Solution

a. Let x represent the number of minutes, and let y represent the charge in dollars. The charge is the monthly fee plus the rate times the number of minutes. Here is the system of equations.

$$\begin{cases} y = 4.80 + 0.05x & \text{Anytime plan} \\ y = 0.09x & \text{TalkMore plan} \end{cases}$$

b. Create a table from the equations. Fill in the times and calculate the charge for each plan. The table shows that when $x = 120$, both y-values are 10.80. Because (120, 10.80) satisfies both equations, it is the solution of the system. The solution means that both plans charge \$10.80 for 120 minutes of long-distance calls.

Long-Distance Plans

Time (min)	Anytime $y = 4.80 + 0.05x$	TalkMore $y = 0.09x$
0	4.80	0
30	6.30	2.70
60	7.80	5.40
90	9.30	8.10
120	10.80	10.80
150	12.30	13.50

On the graph, the solution is the point where the two lines intersect.

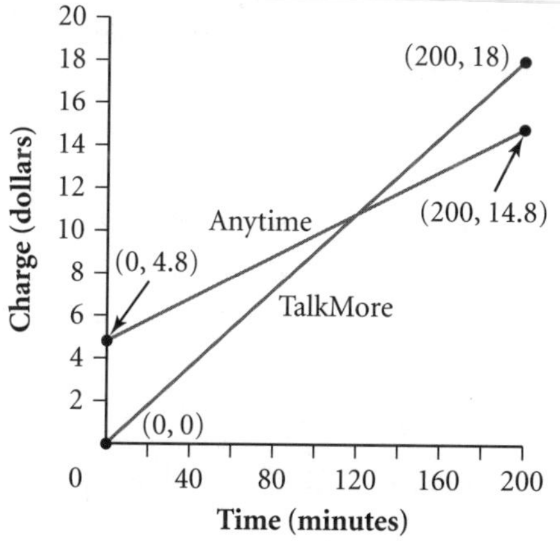

(continued)

Investigation: Where Will They Meet?

Steps 1–4 In this investigation, two students walk along a 6-meter segment. Walker A starts at the 0.5-meter mark and walks toward the 6-meter mark at a rate of 1 m/sec. Walker B starts at the 2-meter mark and walks toward the 6-meter mark at a rate of 0.5 m/sec. Here is a graph of the data collected by one group.

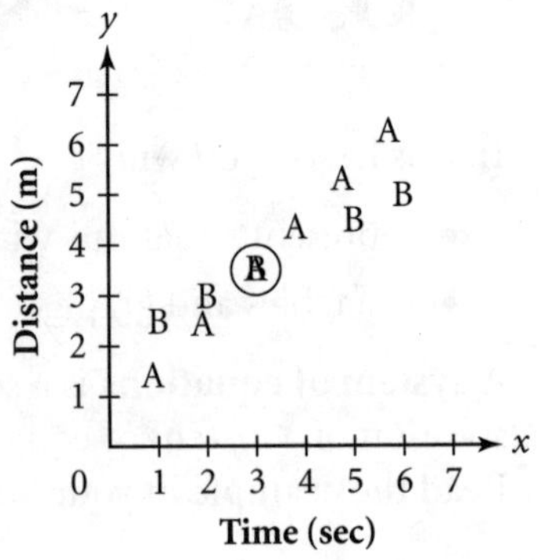

Steps 5–7 You can model this situation with a system of equations and then solve the system to figure out when and where Walker A passes Walker B. If x represents the time in seconds and y represents the distance from the 0-meter mark, the system is

$$\begin{cases} y = 0.5 + x & \text{Walker A} \\ y = 2 + 0.5x & \text{Walker B} \end{cases}$$

Here are graphs of the equations on the same axes. The graphs intersect at (3, 3.5), indicating that Walker A passes Walker B after 3 seconds, when both walkers are at the 3.5-meter mark.

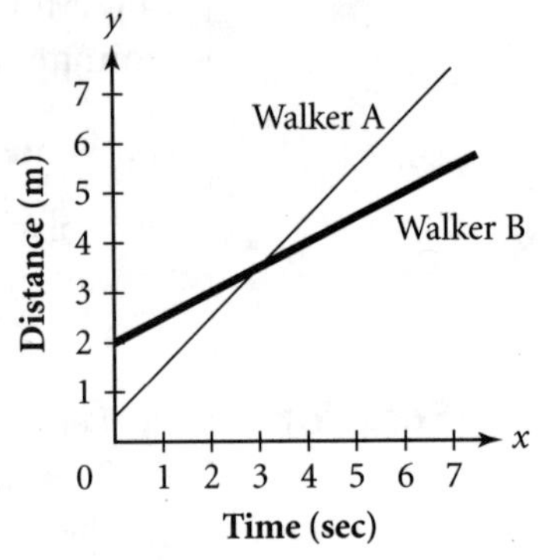

Steps 8–10 If Walker A moved faster than 1 m/sec, the slope of Walker A's line would increase and the intersection point would move closer to the origin, indicating that Walker A passes Walker B sooner and closer to the 0-meter mark.

If the two walkers moved at the same speed, they would never meet. The slope of the lines would be equal, so the lines would be parallel. The system of equations for this situation has no solution.

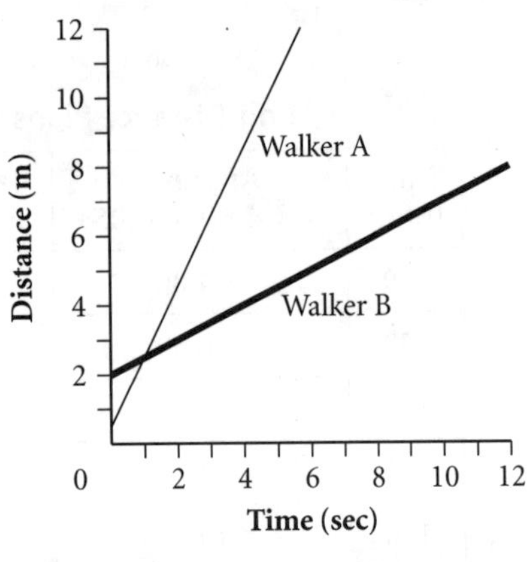

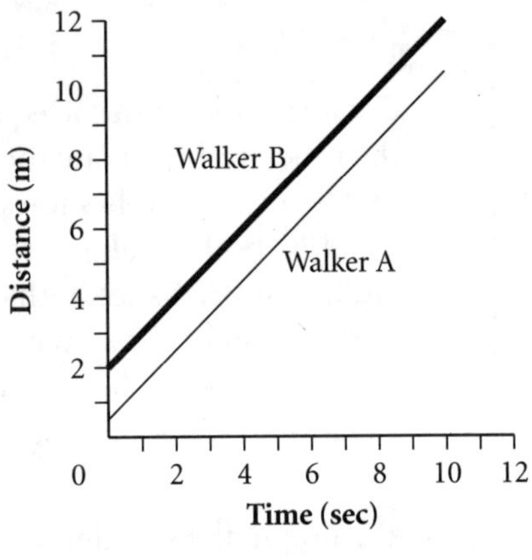

If both walkers walked at the same speed from the same starting mark, the two lines would be identical. Every point on the line is a solution of the system, indicating that the walkers are always at the same location at the same time.

The investigation shows that two lines can intersect at zero points, at one point, or at every point. So a system of linear equations can have zero, one, or an infinite number of solutions.

CONDENSED LESSON

6.2 Solving Systems of Equations Using Substitution

In this lesson you will

- represent situations with **systems of equations**
- use the **substitution method** to solve systems of linear equations

When you use a graph or a table to solve a system of equations, you may only be able to find an approximate solution. The **substitution method** allows you to find an exact solution of a system. Read Example A in your book, which shows how to solve a system using the substitution method.

Investigation: All Tied Up

Start with a thin rope and a thick rope, each 1 meter long. If you tie knots in each rope, measuring the length after each knot, you might get data like this.

Thin Rope	
Number of knots	Length (cm)
0	100
1	94
2	88
3	81.3
4	75.7
5	69.9
6	63.5

Thick Rope	
Number of knots	Length (cm)
0	100
1	89.7
2	78.7
3	68.6
4	57.4
5	47.8
6	38.1

Use the techniques you learned in Chapter 5 to write a linear equation to model the data for each rope.

A possible model for the thin rope is $y = 100 - 6x$, where x is the number of knots and y is the length in centimeters. The y-intercept, 100, is the length of the rope before you tie any knots. The slope, -6, is the change in the length for each knot.

A possible model for the thick rope is $y = 100 - 10.3x$. This equation indicates that the initial length is 100 cm and that the length decreases by 10.3 cm for each knot.

Now, suppose the initial length of the thin rope is 9 meters and the initial length of the thick rope is 10 meters. This system of equations models this situation.

$$\begin{cases} y = 900 - 6x & \text{Length of thin rope} \\ y = 1000 - 10.3x & \text{Length of thick rope} \end{cases}$$

To estimate the solution of this system, make a graph and estimate the point of intersection. The intersection point is about (23, 760).

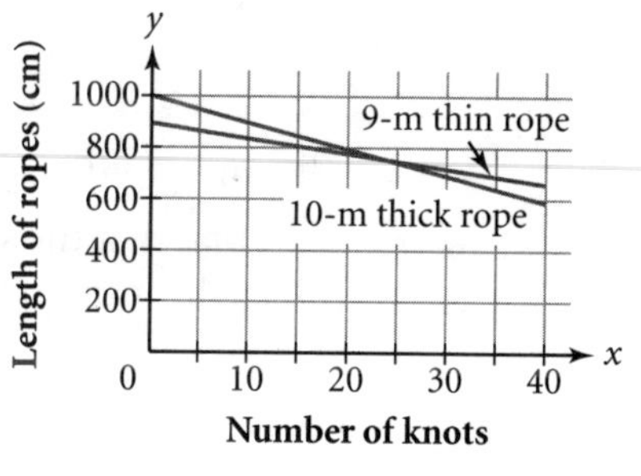

You can also find the solution by using the substitution method. Substitute $900 - 6x$ (from the first equation) for y in the second equation, and solve the resulting equation.

$y = 1000 - 10.3x$	Original second equation.
$900 - 6x = 1000 - 10.3x$	Substitute $900 - 6x$ for y.
$900 = 1000 - 4.3x$	Add $6x$ to both sides and simplify.
$-100 = -4.3x$	Subtract 1000 from both sides.
$23.26 \approx x$	Divide both sides by -4.3.

(continued)

Because x represents the number of knots, the solution must be a whole number. So round x to 23. When x is 23, y is about 760. So the solution is (23, 760). This means that when 23 knots have been tied in each rope, the ropes are about the same length, 760 cm.

Think about how the models would be different if the two ropes had the same thickness. In this situation the slopes would be the same, so the lines would be parallel. In this case the system would have no solutions. In other words, the ropes would never be the same length.

If the ropes had the same thickness *and* the same starting length, the equations and the lines would be exactly the same. In this case there are many solutions. The ropes would be the same length after any number of knots had been tied.

When you solve a system using the substitution method, you sometimes need to rewrite one of the equations before you can substitute. Example B in your book shows you how to solve a system when both equations are given in standard form. Read this example and the text that follows carefully. Then, read the example below.

EXAMPLE

Use the substitution method to solve this system.

$$\begin{cases} 3s = -5t - 3 \\ 2s - 6 = t + 5 \end{cases}$$

▶ ***Solution***

Rewrite one of the equations so that a variable is alone on one side.

$$\begin{cases} 2s - 6 = t + 5 & \text{Original second equation.} \\ 2s - 11 = t & \text{Subtract 5 from both sides.} \end{cases}$$

Now, substitute $2s - 11$ for t in the first equation and solve for s.

$3s = -5(2s - 11) - 3$	Substitute $2s - 11$ for t.
$3s = -10s + 55 - 3$	Distribute the -5.
$3s = -10s + 52$	Subtract 3 from 55.
$13s = 52$	Add $10s$ to both sides.
$s = 4$	Divide both sides by 13.

To find the value of t, substitute 4 for s in either equation and solve for t. You should find that the solution of the system is $(s, t) = (4, -3)$. Check this solution by substituting it into both equations.

CONDENSED LESSON 6.3

Solving Systems of Equations Using Elimination

In this lesson you will

- represent situations with **systems of equations**
- use the **elimination method** to solve systems of linear equations

Read the text at the beginning of Lesson 6.3 in your book. It explains that you can add two equations to get another true equation. Then, read Example A carefully and make sure you understand it. In the example, the variable s is eliminated just by adding the equations. As you will see in the investigation, sometimes using the **elimination method** requires a bit more work.

Investigation: Paper Clips and Pennies

Place one paper clip along the long side of a piece of paper. Then, line up enough pennies to complete the 11-inch length. If you use a jumbo paper clip, you should find that you need 12 pennies.

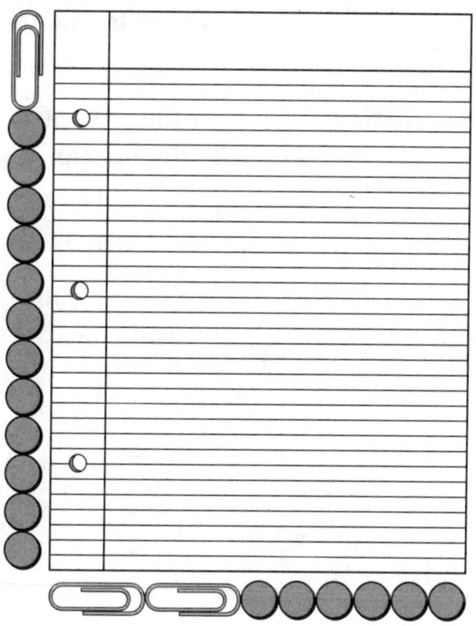

Place two paper clips along the short side of the sheet of paper, and add pennies to complete the 8.5-inch length. With jumbo paper clips, you'll need 6 pennies.

If C is the length of a paper clip and P is the diameter of a penny, you can write this system of equations to represent this situation.

$$\begin{cases} C + 12P = 11 & \text{Long side} \\ 2C + 6P = 8.5 & \text{Short side} \end{cases}$$

Notice that you can't eliminate a variable by adding the two original equations. However, look what happens when you multiply both sides of the first equation by -2.

$$\begin{cases} C + 12P = 11 \\ 2C + 6P = 8.5 \end{cases} \rightarrow \begin{cases} -2C - 24P = -22 \\ 2C + 6P = 8.5 \end{cases}$$

Because you multiplied both sides of the first equation by the same number, the new equation has the same solutions as the original. You can now eliminate the variable C by adding the two equations in the new system.

$$\begin{array}{rl} -2C - 24P = -22 & \\ \underline{2C + 6P = 8.5} & \\ -18P = -13.5 & \text{Add the equations.} \\ P = 0.75 & \text{Divide by } -18. \end{array}$$

To find the value of C, substitute 0.75 for P in either equation and solve for C.

$C + 12(0.75) = 11$ or $2C + 6(0.75) = 8.5$

(continued)

You should find that C is 2. Be sure to check the solution by substituting 0.75 for P and 2 for C in both equations.

$2 + 12(0.75) = 11$ and $2(2) + 6(0.75) = 8.5$

The solution (0.75, 2) means that the penny has a diameter of 0.75 inch and the paper clip has a length of 2 inches.

There are several ways you could have solved the original system of equations. For example, instead of multiplying the first equation by -2, you could have multiplied the second equation by -2. Then, the coefficient of P would be -12 in both equations and you could eliminate P by adding the equations.

Read the rest of the lesson in your book. Here is an additional example.

EXAMPLE At Marli's Discount Music Mart, all CDs are the same price and all cassette tapes are the same price. Rashid bought six CDs and five cassette tapes for \$117.78. Quincy bought four CDs and nine cassette tapes for \$123.74. Write and solve a system of equations to find the price of a CD and the price of a cassette tape.

▶ ***Solution*** If c is the price of a CD and t is the price of a tape, then the problem can be modeled with this system.

$$\begin{cases} 6c + 5t = 117.78 & \text{Rashid's purchase} \\ 4c + 9t = 123.74 & \text{Quincy's purchase} \end{cases}$$

If you multiply the first equation by 2 and the second equation by -3, you will be able to add the equations to eliminate c.

$$\begin{aligned} 6c + 5t = 117.78 \rightarrow \quad 12c + 10t &= 235.56 && \text{Multiply both sides by 2.} \\ 4c + 9t = 123.74 \rightarrow \underline{-12c - 27t} &\underline{= -371.22} && \text{Multiply both sides by } -3. \\ -17t &= -135.66 && \text{Add the equations.} \\ t &= 7.98 && \text{Divide.} \end{aligned}$$

To find the value of c, substitute 7.98 for t in either equation and solve for t.

$$\begin{aligned} 6c + 5t &= 117.78 && \text{Original first equation.} \\ 6c + 5(7.98) &= 117.78 && \text{Substitute 7.98 for } t. \\ 6c + 39.90 &= 117.78 && \text{Multiply.} \\ 6c &= 77.88 && \text{Subtract 39.90 from both sides.} \\ c &= 12.98 && \text{Divide both sides by 6.} \end{aligned}$$

Cassette tapes cost \$7.98 and CDs cost \$12.98. Be sure to check this solution by substituting it into both original equations.

CONDENSED LESSON

6.4

Solving Systems of Equations Using Matrices

In this lesson you will

- represent situations with **systems of equations**
- use **matrices** to solve systems of linear equations

You now know how to solve systems of equations with tables and graphs and by using the substitution and elimination methods. You can also solve systems of equations by using matrices. Pages 331–332 of your book explain how to represent a system of equations with a matrix and then use row operations to find the solution. Read this text and Example A carefully.

Investigation: Diagonalization

Consider this system of equations.

$$\begin{cases} 2x + y = 11 \\ 6x - 5y = 9 \end{cases}$$

Because the equations are in standard form, you can represent the system with a matrix. Write the numerals from the first equation in the first row, and write the numerals from the second equation in the second row.

$$\begin{bmatrix} 2 & 1 & 11 \\ 6 & -5 & 9 \end{bmatrix}$$

To solve the equation, perform row operations to get 1's in the diagonal of the matrix and 0's above and below the diagonal as shown here.

$$\begin{bmatrix} 1 & 0 & a \\ 0 & 1 & b \end{bmatrix}$$

To get a 0 as the first entry in the second row, add -3 times the first row to the second row. This step is similar to using the elimination method to eliminate x from the second equation.

-3 times row 1 →		-6	-3	-33
+ row 2 →	+	6	-5	9
New row 2 →		0	-8	-24

New matrix

$$\begin{bmatrix} 2 & 1 & 11 \\ 0 & -8 & -24 \end{bmatrix}$$

To get 1 as the second entry in the second row, divide that row by -8.

$$\begin{bmatrix} 2 & 1 & 11 \\ 0 & 1 & 3 \end{bmatrix}$$

From the second row, you can see that $y = 3$. Now, subtract the second row from the first to get a 0 as the second entry in the first row. This is similar to substituting 3 for y in the first equation to get $2x = 8$.

Row 1 →		2	1	11
− row 2 →	−	0	1	3
New row 1 →		2	0	8

New matrix

$$\begin{bmatrix} 2 & 0 & 8 \\ 0 & 1 & 3 \end{bmatrix}$$

(continued)

To get a 1 as the first entry in the first row, divide the row by 2.

$$\begin{bmatrix} 1 & 0 & 4 \\ 0 & 1 & 3 \end{bmatrix}$$

You can now see that $x = 4$ and $y = 3$. You can check this solution by substituting it into the original equation.

Example B in your book shows that matrices are useful for solving systems of equations involving large numbers. Here is another example.

EXAMPLE At a college football game, students paid $12 per ticket and nonstudents paid $18 per ticket. The number of students who attended was 1,430 more than the number of nonstudents. The total of all ticket sales was $67,260. How many of the attendees were students, and how many were nonstudents?

▶ ***Solution*** If S is the number of students and N is the number of nonstudents, then you can represent the situation with this system and matrix.

$$\begin{cases} S - N = 1{,}430 \\ 12S + 18N = 67{,}260 \end{cases} \rightarrow \begin{bmatrix} 1 & -1 & 1{,}430 \\ 12 & 18 & 67{,}260 \end{bmatrix}$$

Use row operations to find the solution.

Add -12 times row 1 to row 2 to get new row 2. $\begin{bmatrix} 1 & -1 & 1{,}430 \\ 0 & 30 & 50{,}100 \end{bmatrix}$

Divide row 2 by 30. $\begin{bmatrix} 1 & -1 & 1{,}430 \\ 0 & 1 & 1{,}670 \end{bmatrix}$

Add row 2 to row 1 to get new row 1. $\begin{bmatrix} 1 & 0 & 3{,}100 \\ 0 & 1 & 1{,}670 \end{bmatrix}$

The final matrix shows that $S = 3{,}100$ and $N = 1{,}670$. So 3,100 students and 1,670 nonstudents attended the game. You can check this solution by substituting it into both original equations.

CONDENSED LESSON

6.5 Inequalities in One Variable

In this lesson you will

- write **inequalities** to represent situations
- learn how applying operations to both sides of an inequality affects the direction of the inequality symbol
- solve a problem by writing and **solving an inequality**

An **inequality** is a statement that one quantity is less than or greater than another. Inequalities are written using the symbols $<$, $>$, $\leq$, and $\geq$. Read the text on page 339 of your book, which gives several examples from everyday life and how to write them as inequalities.

Just as with equations, you can solve inequalities by applying the same operations to both sides. However, as you will learn in the investigation, you need to be careful about the direction of the inequality symbol.

Investigation: Toe the Line

In this investigation, two walkers stand on a number line. Walker A starts on the number 2, and Walker B starts on the number 4. You can represent this situation with the inequality $2 < 4$.

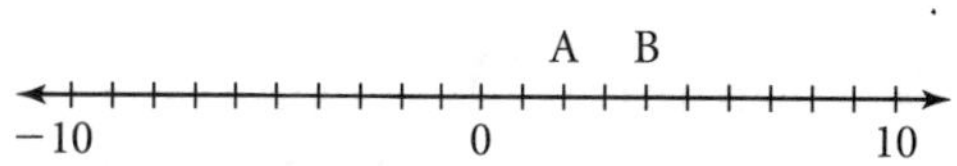

Steps 1–4 When an announcer calls out an operation, the walkers perform the operation on their numbers and move to new positions based on the result. The new positions are represented by an inequality, with the position of Walker A on the left side and the position of Walker B on the right side.

The drawings below show the walkers' positions after the first two operations along with the corresponding inequality.

Operation: Add 2; Inequality: $4 < 6$

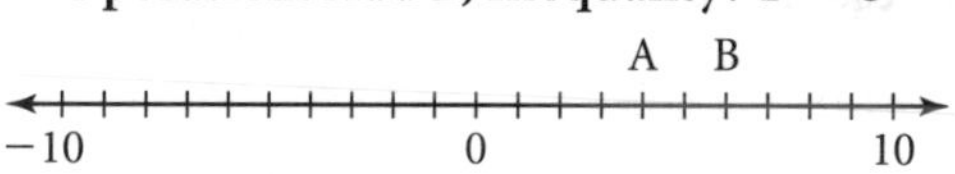

Operation: Subtract 3; Inequality: $1 < 3$

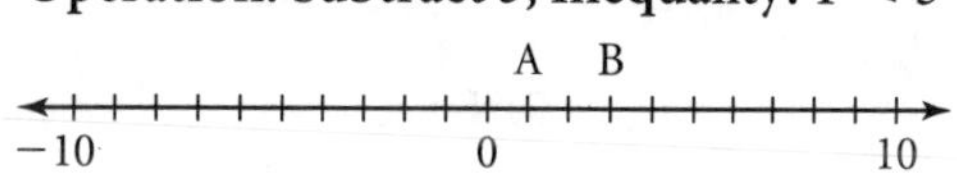

This table shows the results of the remaining operations.

Operation	Walker A's position	Inequality symbol	Walker B's position
Add −2	−1	<	1
Subtract −4	3	<	5
Multiply by 2	6	<	10
Subtract 7	−1	<	3
Multiply by −3	3	>	−9
Add 5	8	>	−4
Divide by −4	−2	<	1
Subtract 2	−4	<	−1
Multiply by −1	4	>	1

(continued)

Steps 5–9 Notice that when a number is added to or subtracted from the walkers' positions, the direction of the inequality (that is, the relative positions of the walkers) remains the same. The direction of an inequality also stays the same when the positions are multiplied or divided by a positive number. However, when the positions are multiplied or divided by a negative number, the direction of the inequality (that is, the relative positions of the walkers) is reversed.

Check these findings by starting with another inequality and applying operations to both sides. You should find that *the direction of the inequality symbol is reversed only when you multiply or divide by a negative number.*

Read Example A in your book, which shows how to graph solutions to inequalities on a number line. Then, read Example B, which applies what you learned in the investigation to solve an inequality. Here is an additional example.

EXAMPLE A Jack takes the bus to the bowling alley. He has \$15 when he arrives. It costs \$2.25 to bowl one game. If Jack needs \$1.50 to take the bus home, how many games can he bowl? Solve this problem by writing and solving an inequality.

▶ ***Solution*** Let g represent the number of games Jack can bowl. We know that the amount Jack starts with minus the amount he spends bowling must be at least (that is, greater than or equal to) \$1.50. So we can write this inequality.

Amount Jack starts with ↘ Cost of bowling g games ↓ Bus fare ↙

$$15 - 2.25g \geq 1.50$$

Now, solve the inequality.

$15 - 2.25g \geq 1.50$	Original inequality.
$15 - 15 - 2.25g \geq 1.50 - 15$	Subtract 15 from both sides.
$-2.25g \geq -13.50$	Subtract.
$\frac{-2.25g}{-2.25} \leq \frac{-13.50}{-2.25}$	Divide both sides by −2.25, and **reverse the inequality symbol.**
$g \leq 6$	Divide.

Jack can bowl 6 games or fewer. Here, $g \leq 6$ is graphed on a number line.

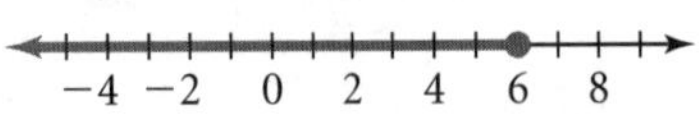

CONDENSED LESSON

6.6 Graphing Inequalities in Two Variables

In this lesson you will

- **graph linear inequalities** in two variables

You know how to graph linear equations in two variables, such as $y = 6 - 3x$. In this lesson you will learn to graph linear inequalities in two variables, such as $y < 6 - 3x$ and $y \geq 6 - 3x$.

Investigation: Graphing Inequalities

To complete this investigation, you'll need a grid worksheet like the one on page 347 of your book.

Choose one of the statements listed on page 347. For each point shown with a circle on the worksheet, substitute the coordinates of the point into the statement, and then fill in the circle with the relational symbol, <, >, or =, that makes the statement true. For example, if you choose the statement $y \square -1 - 2x$, do the following for the point (3, 2):

$y \square -1 - 2x$	Original statement.
$2 \square -1 - 2(3)$	Substitute 3 for x and 2 for y.
$2 \square -7$	Subtract.

Because the symbol > makes this statement true, write > in the circle corresponding to the point (3, 2). Here are completed grids for the four statements.

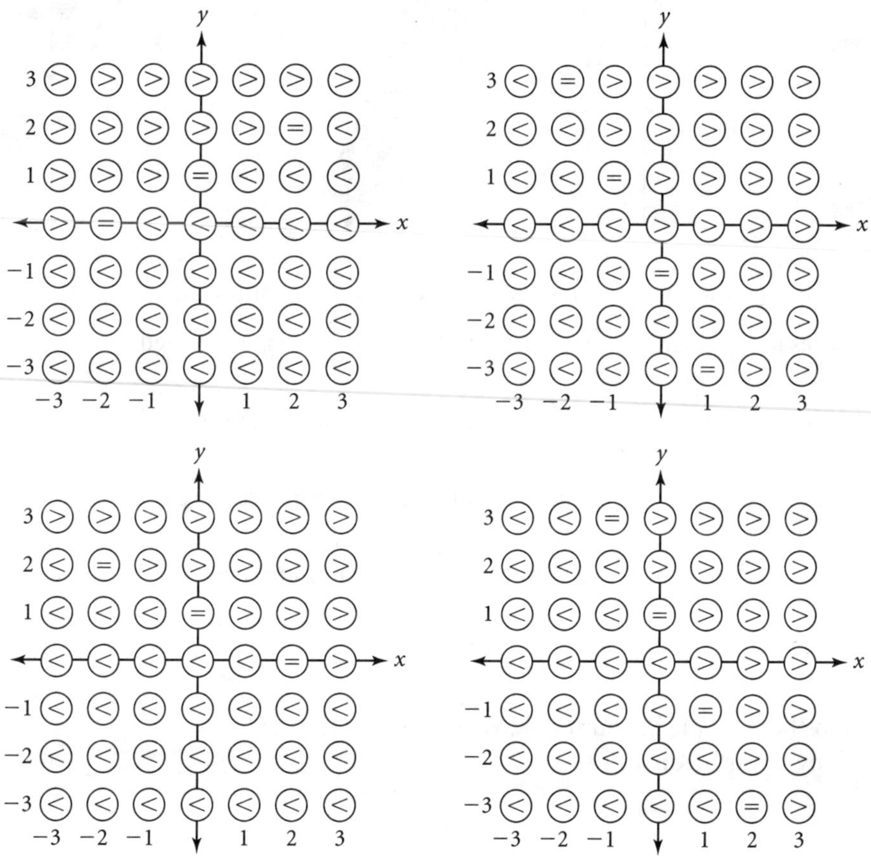

(continued)

Notice that for each statement, the circles containing equal signs form a straight line. All the circles above the line are filled in with $>$ symbols, and all the circles below the line are filled in with $<$ symbols.

Choose one of the statements and test a point with fractional or decimal coordinates. For example, in the grid for $y \square -1 - 2x$, $(-2.2, 1.5)$ is below the line of equal signs. Substitute the coordinates into the statement.

$y \square -1 - 2x$	Original statement.
$1.5 \square -1 - 2(-2.2)$	Substitute -2.2 for x and 1.5 for y.
$1.5 \square 3.4$	Subtract.
$1.5 < 3.4$	Insert the appropriate symbol.

The resulting statement gets a $<$ symbol, just like the other points below the line of equal signs.

Shown here are graphs of $y < -1 - 2x$, $y > -1 - 2x$, $y = -1 - 2x$, $y \leq -1 - 2x$, and $y \geq -1 - 2x$. In each graph the shaded regions include the points that make the statement true. A dashed line indicates that the line is *not* included in the graph. A solid line indicates that the line is included.

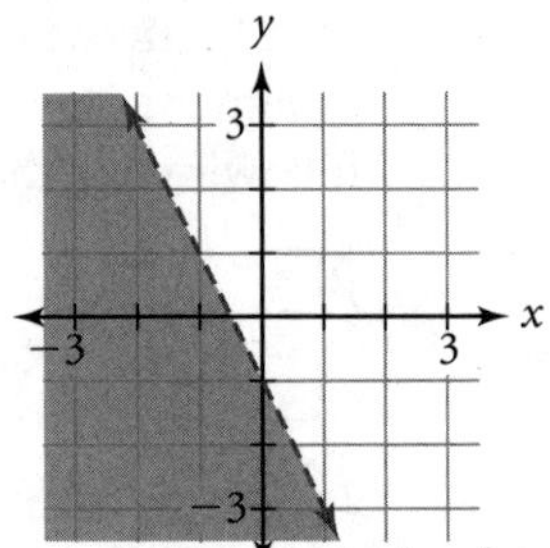

$y < -1 - 2x$

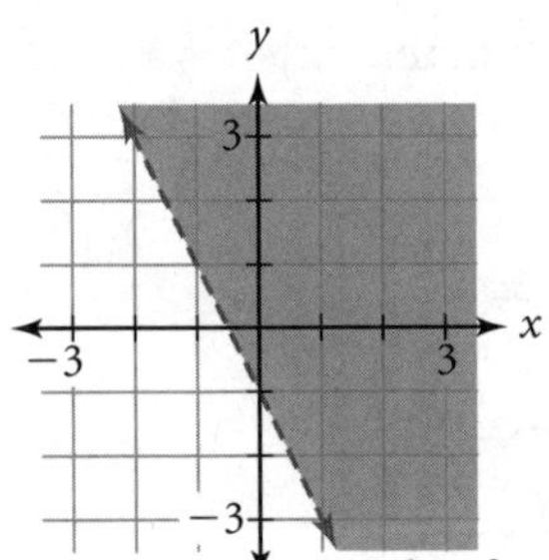

$y > -1 - 2x$

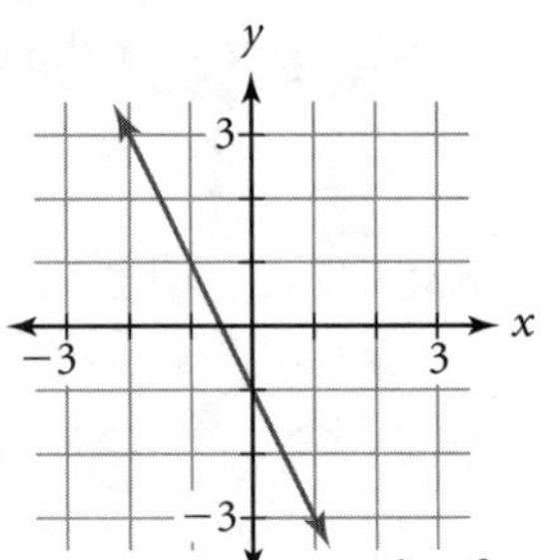

$y = -1 - 2x$

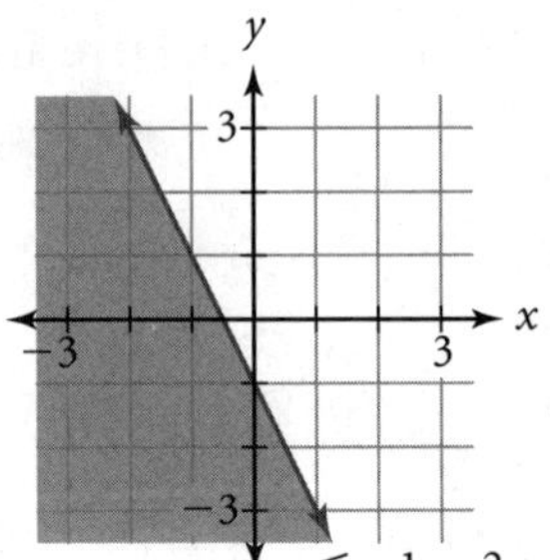

$y \leq -1 - 2x$

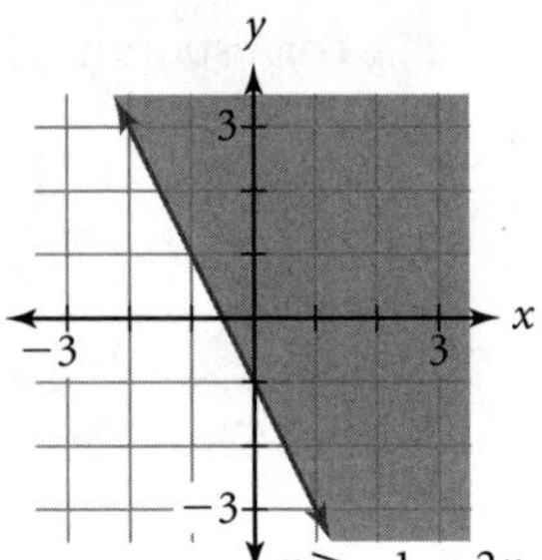

$y \geq -1 - 2x$

Make similar graphs for the other inequalities. You should notice the following:

- Graphs of inequalities in the form $y >$ *expression* and $y \geq$ *expression* are shaded above the line.
- Graphs of inequalities in the form $y <$ *expression* and $y \leq$ *expression* are shaded below the line.
- Graphs of inequalities in the form $y \leq$ *expression* and $y \geq$ *expression* require a solid line.
- Graphs of inequalities in the form $y <$ *expression* and $y >$ *expression* require a dashed line.

Read the rest of the lesson and the example in your book. When you are finished, you should be able to graph any linear inequality.

CONDENSED LESSON

6.7 Systems of Inequalities

In this lesson you will

- **graph solutions** of systems of inequalities
- use systems of inequalities to represent situations involving **constraints**

You can find the solution of a system of equations by graphing the equations and locating the points of intersection. You can use a similar method to find the solution of a system of inequalities. Read Example A in your book. Then, read the additional example below.

EXAMPLE

Graph this system of inequalities and indicate the solution.

$$\begin{cases} y \geq 3 - 2x \\ y < -2 + \frac{3}{4}x \end{cases}$$

▸ Solution

Graph $y = 3 - 2x$ with a solid line because its points satisfy the inequality. Shade above the line because its inequality has the "greater than or equal to" symbol.

Graph $y = -2 + \frac{3}{4}x$ with a dashed line because its points do not satisfy the inequality. Shade below the line because in the inequality y is "less than" the expression in x.

The points in the overlapping region satisfy both inequalities, so the overlapping region is the solution of the system.

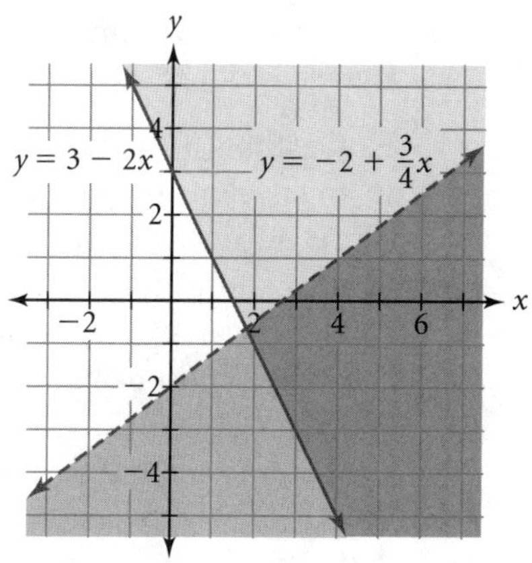

Example B in your book shows how systems of inequalities are useful for modeling situations involving **constraints.** Read through the example.

Investigation: A "Typical" Envelope

Here are two constraints the U.S. Postal Service imposes on envelope sizes.

- The ratio of length to width must be less than or equal to 2.5.
- The ratio of length to width must be greater than or equal to 1.3.

If l and w represent the length and width of an envelope, then the first constraint can be represented by the equation $\frac{l}{w} \leq 2.5$ and the second can be represented by $\frac{l}{w} \geq 1.3$.

You can solve each inequality for l by multiplying both sides by w. This gives the system

$$\begin{cases} l \leq 2.5w \\ l \geq 1.3w \end{cases}$$

(continued)

Note that you do not need to reverse the direction of the inequality symbol when you multiply both sides by y because the width of an envelope must be a positive number.

Here, both inequalities are graphed on the same axes.

The overlap of the shaded regions is the solution of the system. You can check this by choosing a point from the overlapping region and making sure its coordinates satisfy both inequalities.

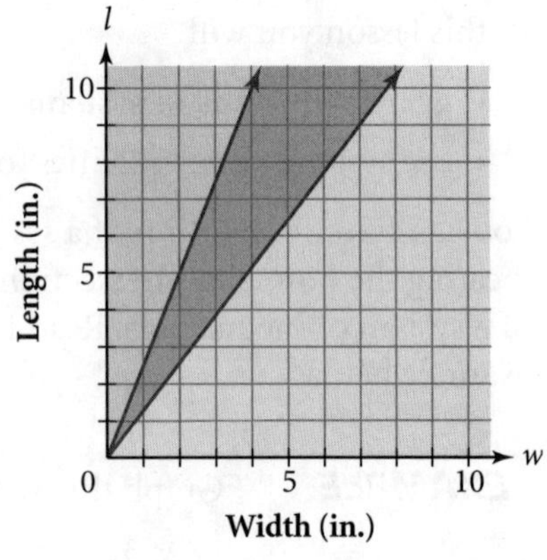

Step 5 in your book gives the dimensions of four envelopes. Points corresponding to these envelopes are plotted on the graph here. Point a, which corresponds to a 5 in.-by-8 in. envelope, and Point d, which corresponds to a 5.5 in.-by-7.5 in. envelope, fall within the overlapping regions, indicating that these envelopes satisfy both constraints.

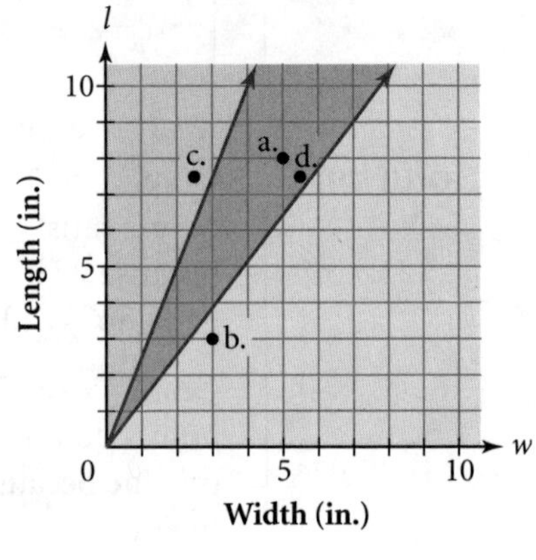

Notice that (0, 0) satisfies the system. This point corresponds to an envelope with no length or width, which does not make sense. Adding constraints specifying minimum and maximum lengths and widths would make the system a more realistic model. For example, for an envelope to require a 34¢ stamp, the length must be between 5 in. and 11.5 in. and the width must be between 3.5 in. and 6.125 in. The system includes these constraints and has this graph.

$$\begin{cases} l \leq 2.5w \\ l \geq 1.3w \\ l \geq 5 \\ w \geq 3.5 \\ l \leq 11.5 \\ w \leq 6.125 \end{cases}$$

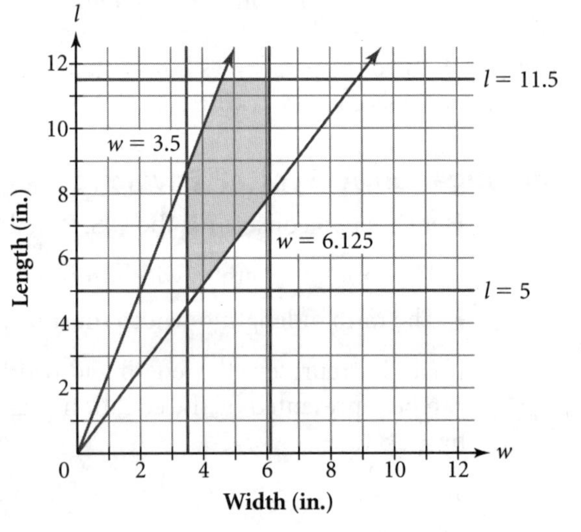

CONDENSED LESSON

7.1 Recursive Routines

In this lesson you will

- explore patterns involving **repeated multiplication**
- write **recursive routines** for situations involving repeated multiplication
- look at **tables and graphs** for situations involving repeated multiplication

In previous chapters you looked at patterns involving repeated addition or subtraction. Such patterns can be modeled with linear equations and straight-line graphs. In this lesson you will begin to explore a different type of pattern.

Investigation: Bugs, Bugs, Everywhere Bugs

Imagine that a bug population starts with 16 bugs and grows by 50% each week. In this investigation you'll look at the pattern of change for this population. In your book, read and follow all the steps of the investigation. Then, look back here to check your results.

This table shows the results for the first 4 weeks.

Bug Invasion

Weeks elapsed	Total number of bugs	Increase in number of bugs (rate of change)	Ratio of this week's total to last week's total
Start (0)	16		
1	24	8	$\frac{24}{16} = \frac{3}{2} = 1.5$
2	36	12	$\frac{36}{24} = \frac{3}{2} = 1.5$
3	54	18	$\frac{54}{36} = \frac{3}{2} = 1.5$
4	81	27	$\frac{81}{54} = \frac{3}{2} = 1.5$

The rate of change for the number of bugs is not constant—it changes from 8 to 12 to 18 to 27—so this pattern is not linear.

Here is a graph of the data. The points have been connected with line segments. Notice that as you move from left to right, the slopes of the line segments increase.

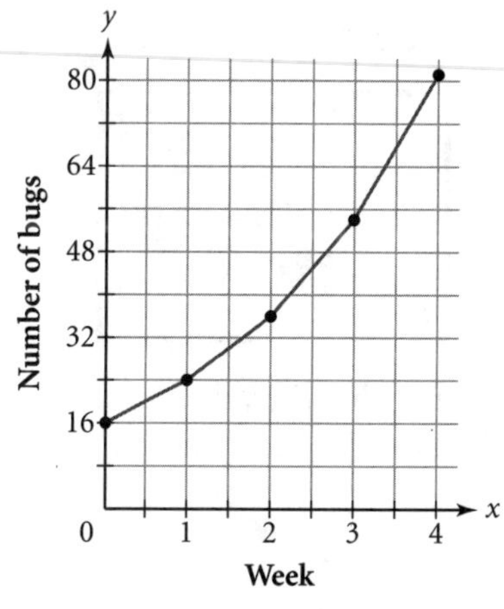

The last column of the table shows that the ratio of the number of bugs each week to the number of bugs the previous week is constant. This constant ratio, 1.5, is the number each week's population is multiplied by to get the next week's population. So, unlike linear patterns, in which you find each value by *adding* a constant number to the previous value, you find each value in this pattern by *multiplying* the previous value by a constant number.

(continued)

You can model the growth in the bug population with this routine.

Press {0, 16} [ENTER]

Press {Ans(1) + 1, Ans(2) * 1.5}

Press [ENTER] to generate each successive term

In the routine, {0, 16} sets the starting bug population (the population for week 0) at 16. The rule {Ans(1) + 1, Ans(2) * 1.5} adds 1 to the week number and multiplies the population by 1.5.

By repeatedly pressing [ENTER], you should find that the populations for weeks 5 through 8 are 122, 182, 273, and 410. The populations for weeks 20 and 30 are 53,204 and 4,602,025.

The example in your book involves compound interest, which shows a pattern of increase involving repeated multiplication. Read the example carefully. Then, read the example below, which involves a decreasing pattern.

EXAMPLE Desi bought a car for \$16,000. He estimates that the value of his car will decrease by 15% a year. What will the car be worth after 4 years? After 7 years?

▶ ***Solution*** Each year the car's value decreases by 15% of its previous value.

Year	Beginning value		Decrease in value		New value
1	16,000	−	16,000 · 0.15	=	16,000(1 − 0.15) or 13,600
2	13,600	−	13,600 · 0.15	=	13,600(1 − 0.15) or 11,560
3	11,560	−	11,560 · 0.15	=	11,560(1 − 0.15) or 9,826

Each year the car's value is multiplied by 1 − 0.15, or 0.85. In other words, the value at the end of each year is 85% of the previous value. You can model this situation with a recursive routine.

Press {0, 16000} [ENTER]

Press {Ans(1) + 1, Ans(2) * 0.85}

Press [ENTER] to generate each successive term

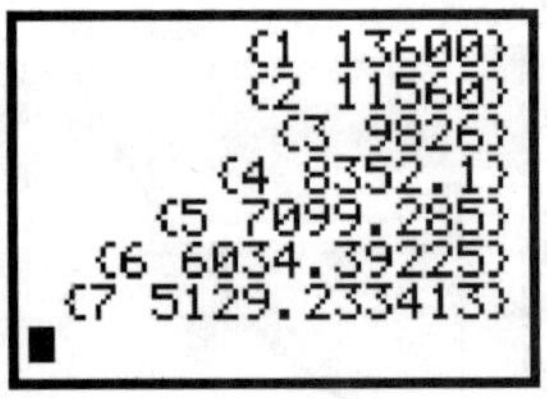

Notice that the value decreases by a smaller amount each year. The value after 4 years is about \$8352. The value after 7 years is about \$5129.

Exponential Equations

In this lesson you will

- write **exponential equations** to represent situations involving a constant multiplier
- change expressions from **expanded form** to **exponential form**
- use exponential equations to model **exponential growth**

You have used recursive routines to generate patterns involving a constant multiplier. In this lesson you'll learn to represent such patterns with equations. This will allow you to find the value of any term without having to find all the terms before it.

Investigation: Growth of the Koch Curve

Here are Stages 0 through 3 of the Koch curve.

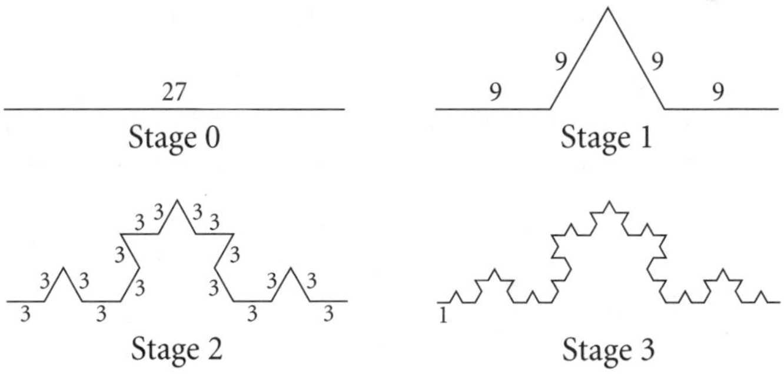

Each stage has four times as many segments as the previous stage, and each segment is $\frac{1}{3}$ the length of the previous segment. This table shows the total length at each stage.

Stage	Total length (units)	Ratio of this stage's length to previous stage's length
0	27	
1	36	$\frac{36}{27} = \frac{4}{3} = 1.\overline{3}$
2	48	$\frac{48}{36} = \frac{4}{3} = 1.\overline{3}$
3	64	$\frac{64}{48} = \frac{4}{3} = 1.\overline{3}$

The ratio of the length at each stage to the length at the previous stage is $\frac{4}{3}$. So the length at each stage is $\frac{4}{3}$ times the length at the previous stage. You can use this fact to find the lengths at Stages 4 and 5.

$$\text{Stage 4 length} = 64 \cdot \frac{4}{3} = 85.\overline{3}$$

$$\text{Stage 5 length} = 85.\overline{3} \cdot \frac{4}{3} = 113.\overline{7}$$

Notice that, at Stage 1, you multiply 27, the original length, by $\frac{4}{3}$ once. At Stage 2, you multiply 27 by $\frac{4}{3}$ twice. At Stage 3, you multiply 27 by $\frac{4}{3}$ three times. You can express this pattern using exponents.

$$\text{Stage 1 length} = 27 \cdot \frac{4}{3} = 27 \cdot \left(\frac{4}{3}\right)^1 = 36$$

$$\text{Stage 2 length} = 27 \cdot \frac{4}{3} \cdot \frac{4}{3} = 27 \cdot \left(\frac{4}{3}\right)^2 = 48$$

$$\text{Stage 3 length} = 27 \cdot \frac{4}{3} \cdot \frac{4}{3} \cdot \frac{4}{3} = 27 \cdot \left(\frac{4}{3}\right)^3 = 64$$

(continued)

In each case the stage number is equal to the exponent. So the length at Stage 5 is $27 \cdot \left(\frac{4}{3}\right)^5 = 113.\overline{7}$, which agrees with the answer on the previous page.

If x is the stage number and y is the total length, then the equation $y = 27 \cdot \left(\frac{4}{3}\right)^x$ models the length at any stage. Here are a calculator graph and a table for this equation.

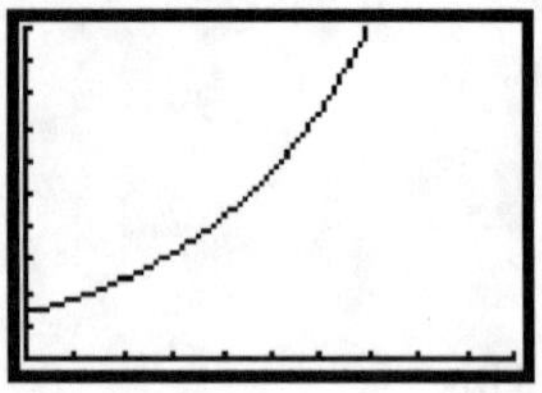

X	Y_1
0	27
1	36
2	48
3	64
4	85.333
5	113.78
6	151.7

X=0

In your book, read the text and examples that follow the investigation. Example A shows how to change expressions from **expanded form** to **exponential form.** Example B explores a situation involving **exponential growth.** Make sure you understand the equation for exponential growth given in the box on page 377. Here is another example.

EXAMPLE Six years ago, Dawn's grandfather gave her a coin collection worth \$350. Since then, the value of the collection has increased by 7% per year. How much is the collection worth now?

▶ ***Solution*** You can model this situation with this equation.

Final value → y; Number of years → x

$$y = 350(1 + 0.07)^x$$

Starting value → 350; Rate of growth → 0.07

To find the current value of the collection—that is, the value 6 years after Dawn received it—substitute 6 for x.

$y = 350(1 + 0.07)^x$ Original equation.

$y = 350(1 + 0.07)^6$ Substitute 6 for x.

$y = 350 \cdot 1.07^6$ Add inside the parentheses.

$y \approx 525.26$ Evaluate the expression.

The collection is now worth \$525.26.

CONDENSED LESSON

7.3 Multiplication and Exponents

In this lesson you will

- use the **multiplication property of exponents** to rewrite expressions
- use the **power properties of exponents** to rewrite expressions

Suppose a savings account starts with a balance of \$500 and earns 3% interest a year. If no money is deposited or withdrawn, the balance after 4 years is $500(1 + 0.03)^4$. Here are two ways you could represent the balance after 5 years.

- You can write $500(1 + 0.03)^5$.
- You can think recursively: The balance after 5 years is the balance after 4 years times the constant multiplier $(1 + 0.03)$. This gives $500(1 + 0.03)^4 \cdot (1 + 0.03)$.

This means that $500(1 + 0.03)^4 \cdot (1 + 0.03) = 500(1 + 0.03)^5$.

In general, you can advance exponential growth by one time period either by multiplying the previous amount by the base or by increasing the exponent by 1. In the investigation you will extend this idea by exploring what happens when you advance by more than one time period.

Investigation: Moving Ahead

Steps 1–3 Look at the expressions in Step 1 in your book. You can write each expression in exponential form with a single base. To see how, first rewrite each expression in expanded form.

a. $3^4 \cdot 3^2 = (3 \cdot 3 \cdot 3 \cdot 3)(3 \cdot 3) = 3^6$

b. $x^3 \cdot x^5 = (x \cdot x \cdot x)(x \cdot x \cdot x \cdot x \cdot x) = x^8$

c. $(1 + 0.05)^2 \cdot (1 + 0.05)^4 = [(1 + 0.05) \cdot (1 + 0.05)]$
$[(1 + 0.05) \cdot (1 + 0.05) \cdot (1 + 0.05) \cdot (1 + 0.05)] = (1 + 0.05)^6$

d. $10^3 \cdot 10^6 = (10 \cdot 10 \cdot 10) \cdot (10 \cdot 10 \cdot 10 \cdot 10 \cdot 10 \cdot 10) = 10^9$

In each case, you add the exponents in the original expression to get the exponent in the final expression. You can generalize these findings as

$b^m \cdot b^n = b^{m+n}$

Steps 4–5 Now complete parts a–c of Step 4 in your book. Here are the answers.

a. If $16(1 + 0.5)^5$ is the number of bugs in the colony after 5 weeks, then $16(1 + 0.5)^5 \cdot (1 + 0.5)^3$ is the number after 3 more weeks (that is, after a total of 8 weeks). This expression can be rewritten as $16(1 + 0.5)^8$.

b. If $11{,}500(1 - 0.2)^7$ is the value of the truck after 7 years, then $11{,}500(1 - 0.2)^7 \cdot (1 - 0.2)^2$ is its value after 2 more years (that is, after a total of 9 years). This expression can be rewritten as $11{,}500(1 - 0.2)^9$.

c. If $A(1 + r)^n$ represents n time periods of exponential growth, then $A(1 + r)^{n+m}$ models m more time periods (that is, a total of $n + m$ time periods).

(continued)

In general, when you are using an exponential model, you can model looking ahead by m time periods by multiplying by $(1 + r)^m$ or by adding m to the exponent.

Page 384 of your book summarizes what you learned in the investigation as the **multiplication property of exponents.** This property is useful for simplifying expressions involving exponents, but keep in mind that it can only be used when the bases are the same. Example A in your book can help you understand why.

Example B illustrates the **power properties of exponents.** Read this example and the following text carefully. Then, read the example below.

EXAMPLE

Use the properties of exponents to rewrite each expression.

a. $(6^4)^3$ **b.** $(3y)^2$ **c.** $5^x \cdot 5^y$

d. $7^3 \cdot 5^2 \cdot 7^1$ **e.** r^4s^4 **f.** $(p^2)^5$

▶ ***Solution***

a. $(6^4)^3 = 6^{4 \cdot 3} = 6^{12}$

b. $(3y)^2 = 3^2y^2$

c. $5^x \cdot 5^y = 5^{x+y}$

d. $7^3 \cdot 5^2 \cdot 7 = 7^{3+1} \cdot 5^2 = 7^4 \cdot 5^2$

e. $r^4s^4 = (rs)^4$

f. $(p^2)^5 = p^{2 \cdot 5} = p^{10}$

CONDENSED LESSON

7.4

Scientific Notation for Large Numbers

In this lesson you will

- write large numbers in **scientific notation**
- convert between scientific notation and standard notation
- use scientific notation to simplify calculations with large numbers

The distance from the Sun to the Andromeda galaxy is 13,000,000,000,000,000,000 miles. In this lesson you will learn about **scientific notation,** a method for writing very large numbers like this in a more compact form. In scientific notation, the distance from the Sun to the Andromeda galaxy is written as 1.3×10^{19} miles.

Investigation: A Scientific Quandary

Steps 1–2 Page 388 of your book gives two lists of numbers. The numbers in the first list are in scientific notation. The numbers in the second list are not. Compare the lists to see if you can figure out what it means for a number to be in scientific notation.

Each of the numbers in scientific notation is written as a product of a number between 1 and 10 and a power of 10. Use this idea to determine which of the numbers in Step 2 are in scientific notation.

a. 4.7×10^3 is in scientific notation.

b. 32×10^5 is *not* in scientific notation because 32 is greater than 10.

c. $2^4 \times 10^6$ is *not* in scientific notation because 2^4 is greater than 10 (and because it is written with an exponent).

d. 1.107×10^{13} is in scientific notation.

e. 0.28×10^{13} is *not* in scientific notation because 0.28 is less than 1.

Steps 3–8 Set your calculator to scientific notation mode. (See **Calculator Note 7C.**) When you enter a number and press ENTER, your calculator will convert the number to scientific notation. Use your calculator to convert 5000 and each of the numbers in Step 5 to scientific notation. Here are the results.

Standard notation	Scientific notation
5,000	5×10^3
250	2.5×10^2
−5,530	-5.53×10^3
14,000	1.4×10^4
7,000,000	7×10^6
18	1.8×10^1
−470,000	-4.7×10^5

(continued)

Notice the following:

- The exponent is equal to the number of digits *after* the first digit in the original number.
- The number multiplied by the power of 10 includes the significant digits of the original number and has one digit to the left of the decimal point.
- If the original number is negative, the number multiplied by the power of 10 is negative.

The exponent 4 is the number of digits after the first digit.

$14{,}000 = 1.4 \times 10^4$

1 and 4 are the significant digits.

To convert 415,000,000 to scientific notation, write 4.15 (the significant digits with one digit to the left of the decimal point). Then, count the number of places you have to move the decimal point to get from 415,000,000 to 4.15. Use the result, 8, as the power of 10. So $415{,}000{,}000 = 4.15 \times 10^8$.

To convert 6.4×10^5 to standard notation, move the decimal point five places to the right adding the zeros you need. So $6.4 \times 10^5 = 640{,}000$.

Read the text and example that follow the investigation in your book. Here is another example.

EXAMPLE Hemoglobin is a protein in red blood cells that transports oxygen from the lungs to the tissues. There are about 25 trillion red blood cells in the average adult human body, and each red blood cell contains 280 million molecules of hemoglobin. How many molecules of hemoglobin does the average adult human body contain?

▶ ***Solution*** First, write the numbers in scientific notation.

25 trillion $= 25{,}000{,}000{,}000{,}000 = 2.5 \times 10^{13}$

280 million $= 280{,}000{,}000 = 2.8 \times 10^8$

Now, multiply the numbers.

$(2.5 \times 10^{13})(2.8 \times 10^8) = 2.5 \times 2.8 \times 10^{13} \times 10^8$	Regroup the numbers.
$= 7.0 \times 10^{13} \times 10^8$	Multiply 2.5 and 2.8.
$= 7.0 \times 10^{21}$	Use the multiplication property of exponents.

There are about 7.0×10^{21} molecules of hemoglobin in the human body.

CONDENSED LESSON

7.5 Looking Back with Exponents

In this lesson you will

- use the **division property of exponents** to rewrite expressions
- relate the division property of exponents to looking *back* in time

You have seen how to multiply expressions with exponents. In this lesson you'll learn how to divide expressions with exponents.

Investigation: The Division Property of Exponents

Steps 1–3 For the expressions in Step 1 in your book, first write the numerators and denominators in expanded form and then eliminate factors equivalent to 1.

a. $\dfrac{5^9}{5^6} = \dfrac{\cancel{5} \cdot \cancel{5} \cdot \cancel{5} \cdot \cancel{5} \cdot \cancel{5} \cdot \cancel{5} \cdot 5 \cdot 5 \cdot 5}{\cancel{5} \cdot \cancel{5} \cdot \cancel{5} \cdot \cancel{5} \cdot \cancel{5} \cdot \cancel{5}} = 5^3$

b. $\dfrac{3^3 \cdot 5^3}{3 \cdot 5^2} = \dfrac{\cancel{3} \cdot 3 \cdot 3 \cdot \cancel{5} \cdot \cancel{5} \cdot 5}{\cancel{3} \cdot \cancel{5} \cdot \cancel{5}} = 3^2 \cdot 5^1$

c. $\dfrac{4^4x^6}{4^2x^3} = \dfrac{\cancel{4} \cdot \cancel{4} \cdot 4 \cdot 4 \cdot \cancel{x} \cdot \cancel{x} \cdot \cancel{x} \cdot x \cdot x \cdot x}{\cancel{4} \cdot \cancel{4} \cdot \cancel{x} \cdot \cancel{x} \cdot \cancel{x}} = 4^2x^3$

Compare the exponents in each final expression to the exponents in the original quotient. Notice that, for each base, the exponent in the final expression is the exponent in the numerator minus the exponent in the denominator. You can use this idea to rewrite the expression in Step 3.

$$\frac{5^{15}\left(1 + \dfrac{0.08}{12}\right)^{24}}{5^{11}\left(1 + \dfrac{0.08}{12}\right)^{18}} = 5^{15-11}\left(1 + \frac{0.08}{12}\right)^{24-18} = 5^4\left(1 + \frac{0.08}{12}\right)^6$$

Steps 4–5 Exponential growth is related to repeated multiplication. When you look ahead in time, you multiply by more constant multipliers. To look back in time, you need to undo some of the multiplication, or divide. Complete parts a–d of Step 4. Here are the answers.

a. If $500(1 + 0.04)^7$ represents the balance after 7 years, then $\frac{500(1 + 0.04)^7}{(1 + 0.04)^3}$ represents the balance 3 years earlier (that is, after 4 years). You can rewrite this expression as $500(1 + 0.04)^4$.

b. If $21{,}300(1 - 0.12)^9$ represents the value after 9 years, then $\frac{21{,}300(1 - 0.12)^9}{(1 - 0.12)^5}$ represents the value 5 years earlier (that is, after 4 years). You can rewrite this expression as $21{,}300(1 - 0.12)^4$.

c. If the population after 5 weeks is $32(1 + 0.50)^5$, then the population 2 weeks earlier was $\frac{32(1 + 0.05)^5}{(1 + 0.50)^2}$ or $32(1 + 0.50)^3$.

d. If $A(1 + r)^n$ models n time periods of exponential growth, then $A(1 + r)^{n-m}$ models the growth m time periods earlier.

In general, to look back m time periods with an exponential growth model, divide by $(1 + r)^m$, where r is the rate of growth, or subtract m from the exponent.

(continued)

In the investigation you explored the **division property of exponents.** Read the statement of the property in your book. Then, read the examples on pages 394 and 395. Here are some more examples.

EXAMPLE A Rewrite each expression with no denominator.

a. $\dfrac{p^7q^5r^3}{p^5q^3r}$ **b.** $\dfrac{5^2 \cdot 2^x \cdot 5^3}{2^y \cdot 5^4}$

▸ ***Solution***

a. $\dfrac{p^7q^5r^3}{p^5q^3r} = p^{7-5}q^{5-3}r^{3-1} = p^2q^2r^2$

b. $\dfrac{5^2 \cdot 2^x \cdot 5^3}{2^y \cdot 5^4} = \dfrac{2^x \cdot 5^{2+3}}{2^y \cdot 5^4} = \dfrac{2^x \cdot 5^5}{2^y \cdot 5^4} = 2^{x-y} \cdot 5^{5-4} = 2^{x-y} \cdot 5$

EXAMPLE B Eight hours ago there were 120 bacteria in a petri dish. Since then the population has increased by 75% each hour.

a. How many bacteria are in the population now?

b. How many bacteria were in the population 5 hours ago?

▸ ***Solution***

a. The population has been increasing for 8 hours. The original population was 120, and the rate of growth is 0.75.

$A(1 + r)^x = 120(1 + 0.75)^8 \approx 10{,}556$

The current population is about 10,556 bacteria.

b. Five hours ago, the population had been growing for 3 hours.

$120(1 + 0.75)^3 \approx 643$

Five hours ago, the bacteria population was about 643.

CONDENSED LESSON

7.6 Zero and Negative Exponents

In this lesson you will

- explore the meaning of **zero and negative exponents**
- rewrite expressions involving negative exponents
- write very small numbers in **scientific notation**

All the exponents you have worked with so far have been positive integers. In this lesson you will explore the meaning of zero and negative-integer exponents.

Investigation: More Exponents

Steps 1–2 Use the division property of exponents to rewrite each of the expressions in Step 1 in your book so that the result has a single exponent.

a. $\frac{y^7}{y^2} = y^5$ **b.** $\frac{3^2}{3^4} = 3^{-2}$ **c.** $\frac{7^4}{7^4} = 7^0$ **d.** $\frac{2}{2^5} = 2^{-4}$ **e.** $\frac{x^3}{x^6} = x^{-3}$

f. $\frac{z^8}{z} = z^7$ **g.** $\frac{2^3}{2^3} = 2^0$ **h.** $\frac{x^5}{x^5} = x^0$ **i.** $\frac{m^6}{m^3} = m^3$ **j.** $\frac{5^3}{5^5} = 5^{-2}$

When the exponent in the numerator is greater than the exponent in the denominator, the result has a positive exponent. When the exponent in the numerator is less than the exponent in the denominator, the result has a negative exponent. When the exponents in the numerator and the denominator are equal, the result has a zero exponent.

Steps 3–6 Look back at all the expressions in Step 1 that gave a result with a negative exponent. You can rewrite these expressions in a different way by first writing them in expanded form and then simplifying.

b. $\frac{3^2}{3^4} = \frac{\cancel{3} \cdot \cancel{3}}{\cancel{3} \cdot \cancel{3} \cdot 3 \cdot 3} = \frac{1}{3^2}$

d. $\frac{2}{2^5} = \frac{\cancel{2}}{\cancel{2} \cdot 2 \cdot 2 \cdot 2 \cdot 2} = \frac{1}{2^4}$

e. $\frac{x^3}{x^6} = \frac{\cancel{x} \cdot \cancel{x} \cdot \cancel{x}}{\cancel{x} \cdot \cancel{x} \cdot \cancel{x} \cdot x \cdot x \cdot x} = \frac{1}{x^3}$

j. $\frac{5^3}{5^5} = \frac{\cancel{5} \cdot \cancel{5} \cdot \cancel{5}}{\cancel{5} \cdot \cancel{5} \cdot \cancel{5} \cdot 5 \cdot 5} = \frac{1}{5^2}$

Compare these results with the results from Step 1. Notice that a base raised to a negative exponent is the same as 1 over the same base raised to the opposite of that exponent.

Now, look back at the expressions in Step 1 that gave a result with a zero exponent. You can rewrite these expressions in a different way by expanding and simplifying.

a. $\frac{7^4}{7^4} = \frac{\cancel{7} \cdot \cancel{7} \cdot \cancel{7} \cdot \cancel{7}}{\cancel{7} \cdot \cancel{7} \cdot \cancel{7} \cdot \cancel{7}} = \frac{1}{1} = 1$

g. $\frac{2^3}{2^3} = \frac{\cancel{2} \cdot \cancel{2} \cdot \cancel{2}}{\cancel{2} \cdot \cancel{2} \cdot \cancel{2}} = \frac{1}{1} = 1$

h. $\frac{x^5}{x^5} = \frac{\cancel{x} \cdot \cancel{x} \cdot \cancel{x} \cdot \cancel{x} \cdot \cancel{x}}{\cancel{x} \cdot \cancel{x} \cdot \cancel{x} \cdot \cancel{x} \cdot \cancel{x}} = \frac{1}{1} = 1$

So a base raised to an exponent of zero is equal to 1.

(continued)

Steps 7–8 Use what you have learned to rewrite each expression in Step 7 in your book so that it has only positive exponents and only one fraction bar.

a. $\frac{5^{-2}}{1} = 5^{-2} = \frac{1}{5^2}$

b. $\frac{1}{3^{-8}} = \frac{1}{\frac{1}{3^8}} = \frac{3^8}{1} = 3^8$

c. $\frac{4x^{-2}}{z^2y^{-5}} = \frac{\frac{4}{x^2}}{\frac{z^2}{y^5}} = \frac{4}{x^2} \cdot \frac{y^5}{z^2} = \frac{4y^5}{x^2z^2}$

As a shortcut, you can rewrite fractions like those above by moving expressions involving exponents from the numerator to the denominator or vice versa, as long as you change the sign of the exponent with each move.

In your book, read the text and examples on pages 400 to 402. Example A gives you more practice simplifying expressions involving exponents. Example B shows how you can use negative exponents to look back in time with exponential growth situations. Example C shows how scientific notation can be used to write very small numbers. Below is an additional example involving scientific notation.

EXAMPLE Convert each number from standard notation to scientific notation or vice versa.

a. An angstrom is a tiny unit of length equal to about 0.000000003973 inch.

b. A proton has a mass of about 1.67×10^{-24} grams.

▶ ***Solution***

a. $0.000000003973 = 3.973 \times 0.000000001$

$$= 3.973 \times \frac{1}{1{,}000{,}000{,}000}$$

$$= 3.973 \times \frac{1}{10^9}$$

$$= 3.973 \times 10^{-9}$$

In general, to rewrite a number less than 1 in scientific notation, count the number of places you must move the decimal to the right to get a number between 1 and 10. Use the negative of that number as the exponent of 10.

b. $1.67 \times 10^{-24} = \frac{1.67}{10^{24}}$

$$1= \frac{1.67}{1{,}000{,}000{,}000{,}000{,}000{,}000{,}000{,}000}$$

$$1= 0.00000000000000000000000167$$

In general, when you are given a number in scientific notation with a negative exponent, you can convert it to standard notation by moving the decimal point to the left the number of places indicated by the exponent.

CONDENSED LESSON 7.7

Fitting Exponential Models to Data

In this lesson you will

- fit **exponential models** to data
- use exponential models to **make predictions**

In earlier chapters you wrote equations to model linear data. In this lesson you will write equations to model data that shows an exponential pattern of growth or decay.

Investigation: Radioactive Decay

This investigation models the radioactive decay of a substance. Below, we work with sample data collected in one classroom. However, if you have the materials, it is a good idea to try collecting the data and completing the investigation on your own before reading the text below.

Steps 1–3 Read through Steps 1–3 in your book. Here is the data collected by one group. The last column will be discussed later.

"Years" elapsed	"Atoms" remaining	Successive ratios
0	201	
1	147	0.7313
2	120	0.8163
3	94	0.7833
4	71	0.7553
5	52	0.7324
6	42	0.8077
7	32	0.7619
8	28	0.8750
9	22	0.7857
10	18	0.8182
11	15	0.8333
12	12	0.8000
13	10	0.8333
14	9	0.9000

Steps 4–10 Here is a scatter plot of the data. Notice that the points appear to follow an exponential pattern.

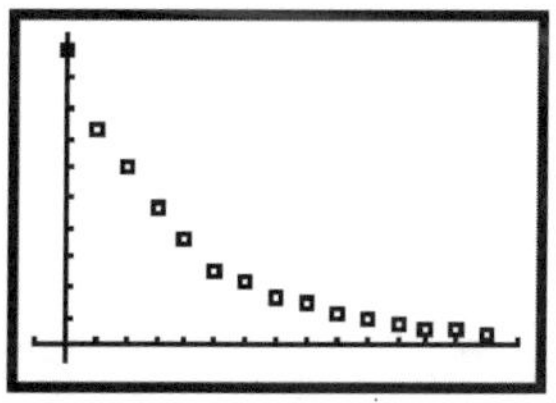

[0, 14, 1, 0, 200, 25]

To fit an exponential equation to this data, you need to find a number to use as the constant multiplier. To do this, first compute the ratios of successive "atoms remaining" values. The results are shown in the table. The ratios are fairly close. We'll use the mean, 0.802, as a representative ratio. Because about 0.802 or 80.2% of the atoms remain each year, about 100% − 80.2%, or 19.8%, of the atoms decay.

(continued)

So to write an exponential equation to model this situation, we can use starting value 201 (the number of "atoms" at the start of the experiment) and decay rate 0.198. The equation is $y = 201(1 - 0.198)^x$. Here, this equation is graphed in the same window as the scatter plot.

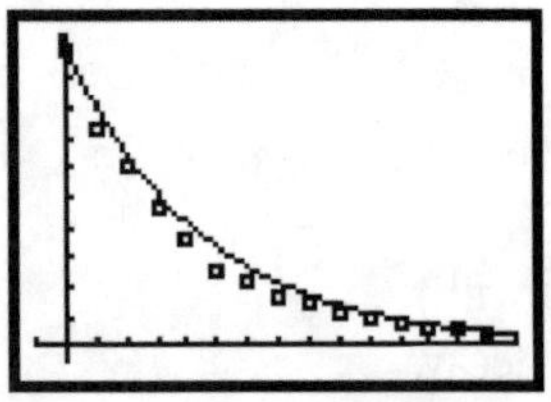

The equation does not appear to fit the data very well. Adjust the values of A and r until you get a better fit.

Here is the graph of $y = 195(1 - 0.220)^x$. This equation appears to fit the data quite well.

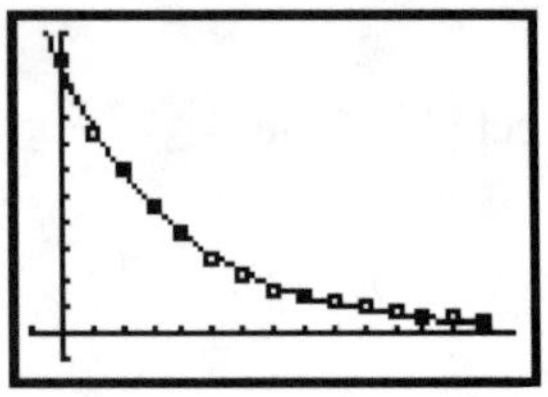

Steps 11–12 The group that collected the data in the table used a plate with a 68° angle. The section enclosed by the angle made up $\frac{68}{360}$ or 19% of the plate's area. This is close to the decay rate used in the equation model. This makes sense because if the counters land so that they are distributed evenly, about 19% of them will land in the 68° section—that is, 19% of the atoms will decay.

Create a calculator table for the model equation $y = 195(1 - 0.220)^x$. Compare the values in the calculator table to those in the table of data. Notice that while none of the values in the calculator table are exactly the same as the actual data values, most of them are very close.

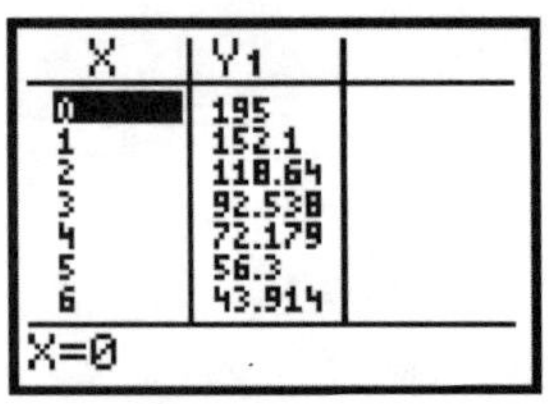

X	Y1
0	195
1	152.1
2	118.64
3	92.538
4	72.179
5	56.3
6	43.914

X=0

Now, read the text and example that follow the investigation in your book.

CONDENSED LESSON

8.1 Secret Codes

In this lesson you will

- use a **coding grid** to write a coded message
- create and use a **letter-shift code**
- determine whether given relationships are **functions**

You have studied many relationships between variables. In this lesson you will learn about a special type of relationship called a **function.**

Investigation: TFDSFU DPEFT

The table and grid on pages 424 and 425 of your book represent a letter-shift code. Read the text before Step 1, which explains how to use the code to write messages.

Steps 1–3 Use the coding grid to write a short coded message. For example, JHO TUSETYDW JXYI is the code for TRY DECODING THIS.

If you think of the letters A through Z as corresponding to the numbers 1 through 26, you can use either of the following rules to code a letter:

- Add 16 to the letter's number. If the result is less than 26, record the corresponding letter. If the result is greater than 26, subtract 26, and record the letter that corresponds to the result.
- Subtract 10 from the letter's number. If the result is greater than 1, record the corresponding letter. If the result is less than 1, add 26, and record the letter that corresponds to the result.

Steps 4–7 Now, create your own code by writing a rule that shifts the letters a specified number of places. Then, put your code on a grid. The grid shows a code in which letters are shifted by 5 letters.

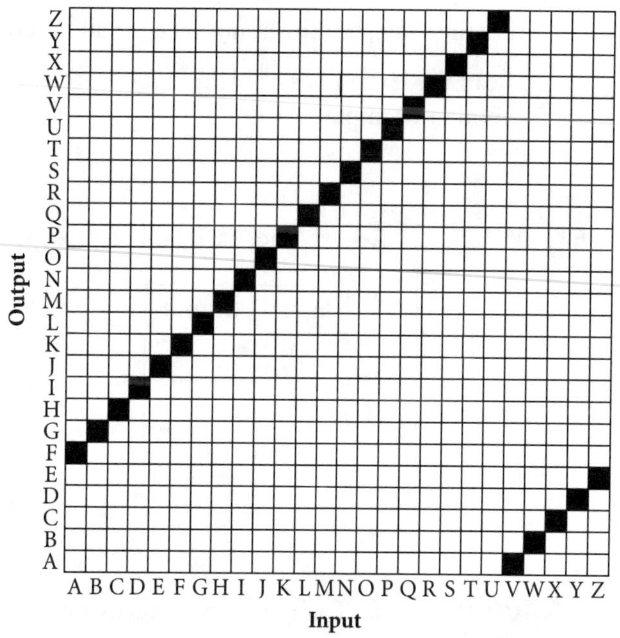

Using this grid, the message TRY DECODING THIS is coded YWD IJHTISL YMNX.

For any letter-shift code, the grid will show two parallel lines of shaded squares, with no row or column containing more than one shaded square.

Steps 8–12 Page 426 shows a grid for a different code. Try using the grid to decode this word: MHJLH.

The coded word is WATER. Were you able to decode it? You may have had trouble because some of the code letters could represent two possible letters in the actual word. For example, H could represent A or R, and J could represent C or T.

Now, use the grid to code the word FUNCTION. You should find that there are

(continued)

several possible codes. The grid indicates that every letter between K and S can be coded in two different ways.

The grid on page 425 makes it much easier to code and decode messages because each input letter corresponds to exactly one output letter.

Use a coding grid to create a coding scheme in which each input letter corresponds to one output letter and no two shaded squares touch one another. The grid shows one possibility.

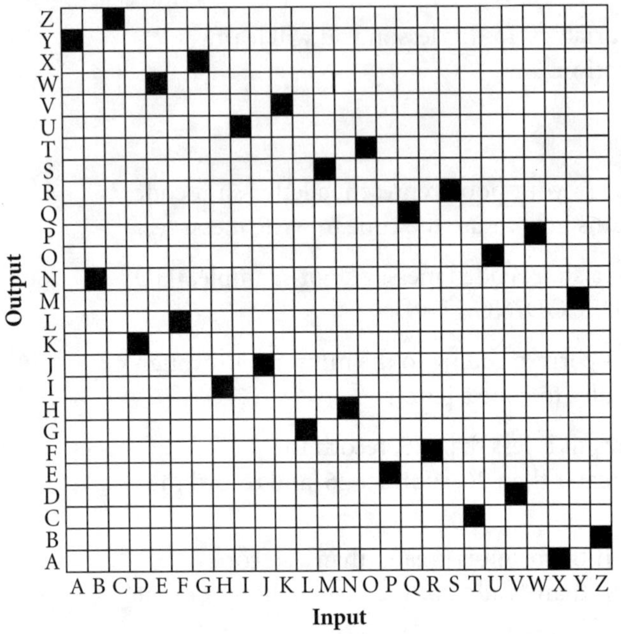

The codes you looked at in the investigation are relationships. A relationship, such as a letter-shift code, for which there is exactly one output for each input is called a **function.** The set of input values for a function is the **domain** of the function, and the set of output values is the **range.** In the letter-shift code, the domain and range are the same, but this is not true for all functions.

Read the text and example that follow the investigation in your book. Here is another example.

EXAMPLE Tell whether each table represents a function.

a.

Input	Output
1	4
2	3
3	4
4	3

b.

Input	Output
red	rose
blue	sky
yellow	sun
blue	ocean

c.

Input	Output
A	a
B	b
C	c
D	d

▸ ***Solution***

a. Each input has exactly one output, so this is a function.

b. The input blue has two outputs (sky and ocean), so this is not a function.

c. Each input has exactly one output, so this is a function.

CONDENSED LESSON 8.2

Functions and Graphs

In this lesson you will

- represent relationships with tables, graphs, and equations
- use the **vertical line test** to determine whether a relationship is a function

You have written and used many rules that transform one number into another. For example, one simple rule is "Multiply each number by 2." You can represent this rule with a table, an equation, a graph, or a diagram.

Table

Input x	Output y
3	6
−14	−28
9.3	18.6
x	$2x$

Equation

$y = 2x$

Graph

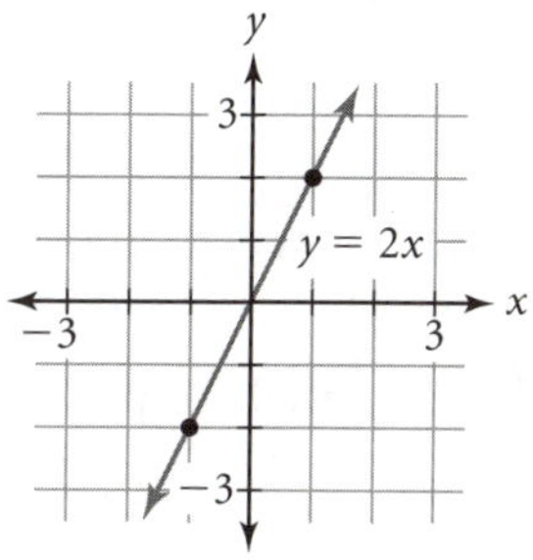

Diagram

Domain Range

3 → 6

−14 → −28

9.3 → 18.6

In this lesson you will learn a method for determining whether a rule is a function based on its graph.

Investigation: Testing for Functions

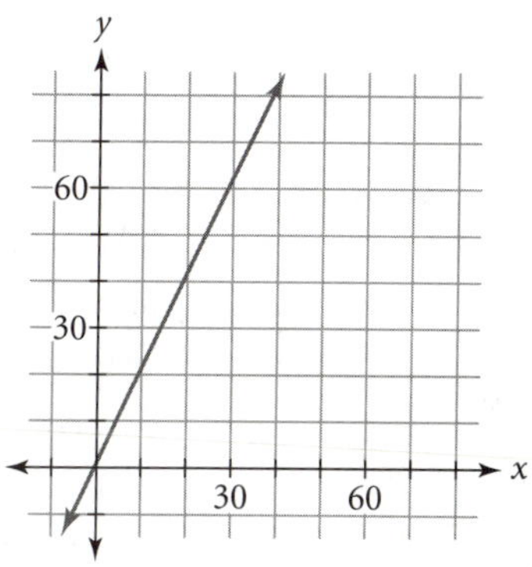

Page 433 shows three relationships, each in a different form. First, we'll look at Relationship A, which is given as a table. Notice that each output in the table is one more than twice the input. You can represent this relationship with the equation $y = 2x + 1$ and with a graph.

Both the domain and the range of the relationship $y = 2x + 1$ are the set of all numbers.

Move a vertical line, such as the edge of a ruler, from side to side on the graph. Notice that in any position the line intersects the graph only once. This means that there is only one output value for each input value, so the relationship is a function.

Now, consider Relationship B. Here is a possible table of values and a graph of the relationship.

You can substitute any value for x in $y = 47(1.10)^x$, but the result will always be greater than 0. So the domain is all numbers, and the range is all numbers greater than 0. Use the **vertical line test** to determine whether this relationship is a function. You should find that any vertical line intersects the graph only once, indicating that the relationship is a function.

x	y
−15	11.25
−10	18.12
−5	29.18
0	47
5	75.69
10	121.91
15	196.33

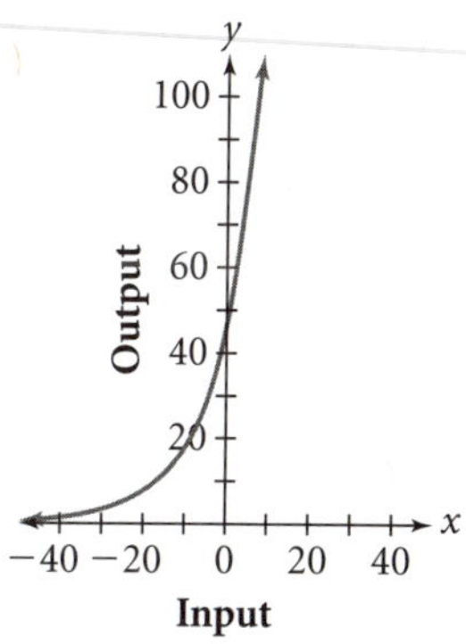

(continued)

Relationship C is given as a graph. You can represent the graph by the inequality $y < -1 + \frac{2}{3}x$. Here is a possible table of values for the relationship.

x	y
−4	−5
−1	−4
0	−3
3	0
6	2

Both the domain and the range of this relationship are the set of all numbers. If you use the vertical line test on the graph, you'll find that any vertical line intersects the graph at more than one point (in fact, any line crosses the graph at an infinite number of points). So Relationship C is not a function.

The vertical line test helps you determine whether a relationship is a function by looking at its graph. If all possible lines cross the graph only once or not at all, then the graph is a function. If even one vertical line crosses the graph more than once, then the graph is not a function.

Read the rest of the lesson in your book. Then, read the example below.

EXAMPLE Use the vertical line test to determine which relationships are functions.

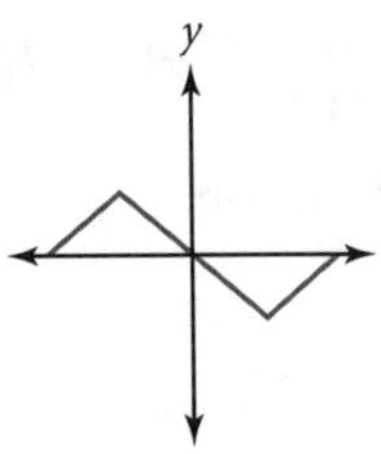

Relationship A

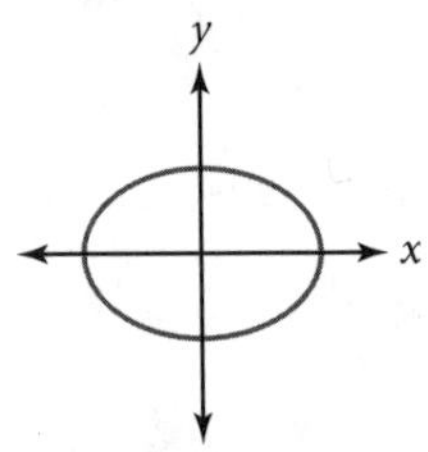

Relationship B

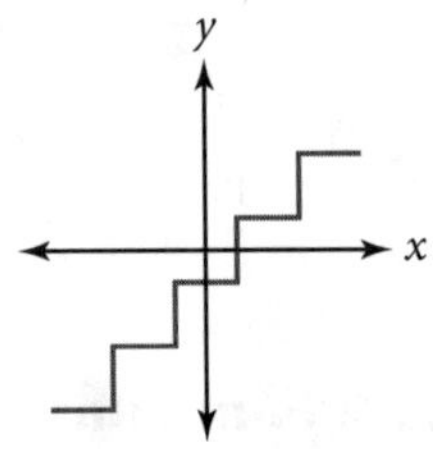

Relationship C

▸ ***Solution*** Relationship A is a function because any vertical line crosses the graph only once. Relationship B is not a function because you can draw a vertical line that crosses the graph twice. Relationship C is not a function because any vertical line through a vertical segment of the graph crosses the graph more than once.

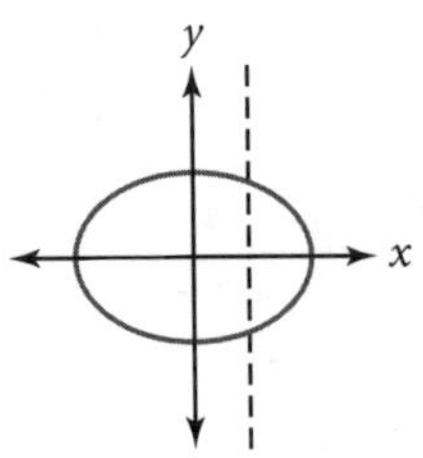

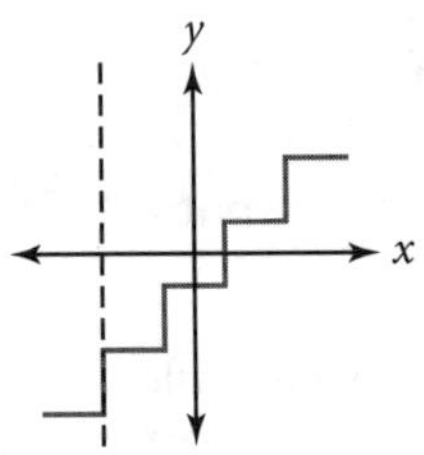

CONDENSED LESSON

8.3 Graphs of Real-World Situations

In this lesson you will

- write a description of a real-world relationship displayed in a graph
- draw a graph to match a description of a real-world situation

In this lesson you'll look at graphs that show how two real-world quantities are related. The example on page 440 of your book shows the relationship between time and the depth of water in a leaky swimming pool. Read the example carefully, and make sure you understand how the description of the situation fits the graph. Here is another example.

EXAMPLE

This graph shows the volume of air in a balloon as it changes over time. Tell what quantities are varying and how they are related. Give possible real-world events in your explanation.

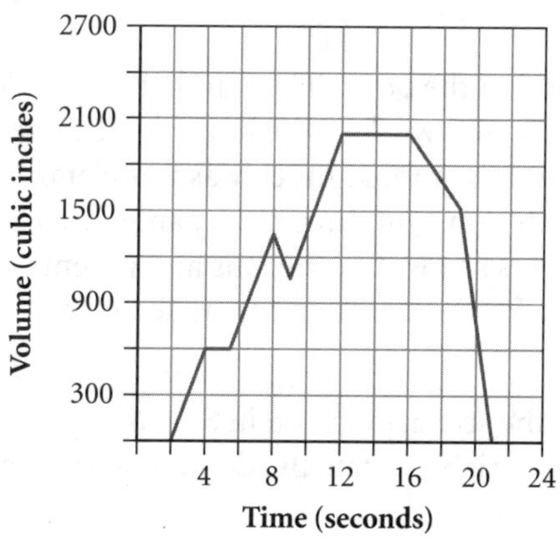

Solution

The graph shows how the volume of air changes over time. The balloon is completely deflated for about the first 2 seconds, that is, for $0 \leq t \leq 2$. From 2 seconds to about 4 seconds ($2 \leq t \leq 4$), the balloon is inflated at a fairly steady rate. Between the 4- and 5.5-second marks, the volume stays constant at about 600 cubic inches. Perhaps the person blowing up the balloon is holding the balloon closed while she takes a breath. In the period $5.5 \leq t \leq 8$, the balloon is being inflated again. From $t = 8$ to about $t = 9$, the volume of air decreases slightly. The person might be taking another break but not holding the balloon closed tightly, allowing air to escape. During the period $9 \leq t \leq 12$, the balloon is inflated more. Then, from $t = 12$ to about $t = 16$, the volume is steady at about 2000 cubic inches. Perhaps the balloon is fully inflated so that the person stops blowing and holds the balloon closed. From about $t = 16$ to $t = 19$, the balloon is deflating slowly. The person may be holding the balloon partially closed. Between $t = 19$ and $t = 22$, the balloon is deflating quickly. The person might have let go of the balloon, allowing the air to escape rapidly.

In the example above, the volume of air is a function of time. Because the volume of air *depends* on the time, it is the **dependent variable.** Time is the **independent variable.** Read the discussion of dependent and independent variables and domain and range on page 441 of your book.

(continued)

Investigation: Identify the Variables

Read the story about the volleyball game at the beginning of the investigation. In this situation, the independent variable is the time in hours. The values in the domain are positive numbers. It might take about 6 hours for the pre-game warm-ups, the game, the dance, and the cleanup, so a reasonable domain would be $0 \leq x \leq 6$.

The dependent variable in this situation is the number of people, which must be a whole number. The range might go from about 10 (the number of people left in the gym after the dance) to several hundred (including the players, coaches, fans, and workers).

Now, sketch a graph that matches the story. Here is one possible graph.

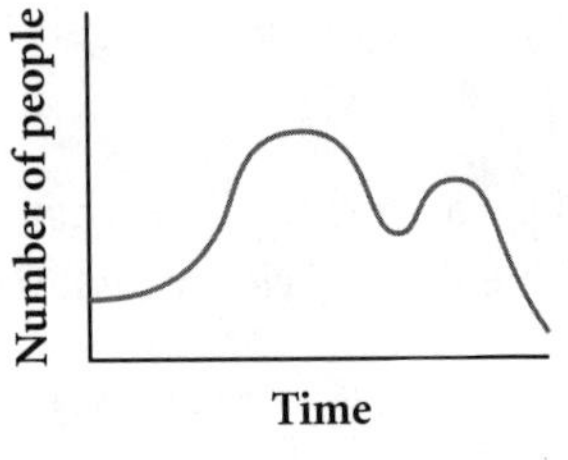

Think about how this graph fits the description in the book. The number of people starts out small (when only the coaches, players, and workers are in the gym). The number then starts increasing slowly at first and then more quickly as more and more fans arrive. The number of people is fairly constant during the game. After the game is over, the number of people decreases as most of the fans and parents leave. The number increases as students arrive for the dance and then decreases after the dance.

Now, create your own graph based on one of the scenarios given in Step 6 in your book. Define the dependent and independent variables, state the domain and range, and then sketch a graph.

Read the text following the investigation in your book.

CONDENSED LESSON

8.4 Function Notation

In this lesson you will

- learn to use **function notation**
- use a graph to evaluate a function for various input values
- use an equation to evaluate a function for various input values

The equation $y = 1 - 2x$ represents a function. You can use the letter f to name this function and then use **function notation** to express it as $f(x) = 1 - 2x$. You read $f(x)$ as "f of x," which means "the output value of the function f for the input value x." So, for example, $f(2)$ is the value of $1 - 2x$ when x is 2, so $f(2) = -3$. (Note: In function notation, the parentheses do *not* mean multiplication.)

Not all functions are expressed as equations. Here is a graph of a function g. The equation is not given, but you can still use function notation to express the outputs for various inputs. For example, $g(0) = 3$, $g(4) = 6$, and $g(6) = 1$. Can you find x-values for which $g(x) = 3$?

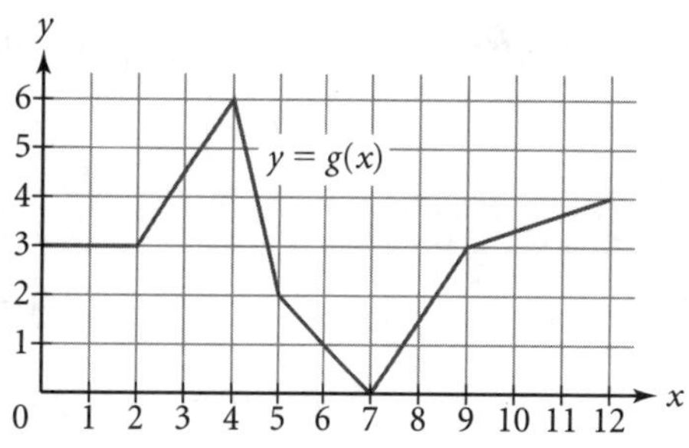

Investigation: A Graphic Message

Look at the graph given at the beginning of the investigation. The domain of this function is $0 \le x \le 26$, and the range is $0 \le y \le 20$.

Look at the sequence given in Step 2 in your book. Use the graph of f to evaluate the function for the given inputs, and find the value of each term of the sequence. Here are the results.

$f(3) = 5$

$f(18) + f(3) = 16 + 5 = 21$

$f(5) \cdot f(4) = 3 \cdot 4 = 12$

$f(15) \div f(6) = 10 \div 2 = 5$

$f(20) - f(10) = 20 - 2 = 18$

So the sequence is 5, 21, 12, 5, 8.

Now, look at the sequence given in Step 3. Evaluate f for the given inputs, and find the value of each term of the sequence. Here are the results.

$f(0) + f(1) - 3 = 8 + 7 - 3 = 12$

$5 \cdot f(9) = 5 \cdot 1 = 5$

When $f(x) = 10$, $x = 15$

$f(9 + 8) = f(17) = 14$

$$\frac{f(17) + f(10)}{2} = \frac{14 + 2}{2} = 8$$

$f(8 \cdot 3) - 5 \cdot f(11) = f(24) - 5 \cdot 3 = 16 - 15 = 1$

$f(4 \cdot 5 - 1) = f(19) = 18$

$f(12) = 4$

So the sequence is 12, 5, 15, 14, 8, 1, 18, 4.

(continued)

Now, think of the numbers 1 through 26 as the letters A through Z. Replace each number in the sequences you found in Steps 2 and 3 with the corresponding letter. The sequence in Step 2 gives EULER, and the sequence in Step 3 gives LEONHARD. Leonhard Euler (pronounced "oiler") invented much of the mathematical notation in use today.

EXAMPLE

You can use the function $f(x) = -31.25 + 1.35x$ to approximate the wind chill temperature $f(x)$ for a given actual temperature when the wind speed is 15 miles per hour. Both x and $f(x)$ are in degrees Fahrenheit. Find $f(x)$ for each given value of x.

a. $f(-10)$ **b.** $f(0)$ **c.** $f(30)$ **d.** $f(5)$

▶ ***Solution***

In each case, substitute the value in parentheses for x in the function.

a. $f(-10) = -31.25 + 1.35(-10)$

$= -31.25 + (-13.5)$

$= -44.75$

b. $f(0) = -31.25 + 1.35(0)$

$= -31.25 + 0$

$= -31.25$

c. $f(30) = -31.25 + 1.35(30)$

$= -31.25 + 40.5$

$= 9.25$

d. $f(5) = -31.25 + 1.35(5)$

$= -31.25 + 6.75$

$= -24.5$

(See **Calculator Note 8A** to learn how to evaluate functions on your calculator.) Your calculator uses the notation $Y_1(x)$ instead of $f(x)$. The function is the equation you have entered into Y_1. When you write an equation for a function, you can use any letters you want to represent the variables and the function. For example, you might use $W(t) = -31.25 + 1.35t$ for the wind chill function discussed above.

CONDENSED
LESSON

8.5 Interpreting Graphs

In this lesson you will

- describe graphs using the words **increasing, decreasing, linear,** and **nonlinear**
- match graphs with descriptions of real-world situations
- learn about **continuous** and **discrete** functions
- use intervals of the domain to help you describe a function's behavior

You have seen that graphs are useful for showing how quantities are related. In this lesson you'll practice interpreting and describing graphs. Pages 452–453 of your book discuss graphs of linear and nonlinear functions. Read this text carefully, and make sure you understand it.

Investigation: Matching Up

All the graphs in Step 1 of the investigation show **increasing** functions, meaning that as the x-values increase, the y-values also increase. In Graph A, the function values increase at a constant rate. In Graph B, the values increase slowly at first and then more quickly. In Graph C, the function switches from one constant rate of increase to another.

The graphs in Step 2 show **decreasing** functions, meaning that as the x-values increase, the y-values decrease. In Graph D, the function values decrease at a constant rate. In Graph E, the values decrease quickly at first and then more slowly. In Graph F, the function switches from one constant rate of decrease to another.

Now, look at the graphs on page 454 of the investigation. Then, read the description of Situation A. The independent variable for this situation is time, and the dependent variable is the number of deer. Think about how the population changes over time, and make a sketch. Your sketch should look somewhat like Graph 5, which is a nonlinear graph at first and then increases at an increasing rate. Then, the rate of change slows down and the graph becomes nearly linear with a very small rate of change. This fits the description of Situation A, which states that the number of deer initially increased by a steady *percentage* (indicating exponential growth) and then the growth rate leveled off.

Read and make a sketch for Situation B. The independent variable for this case is time in days, and the dependent variable is hours of daylight. Your sketch should look somewhat like Graph 3, which is a nonlinear graph that increases slowly at first, then increases more quickly, then levels off and peaks, then decreases quickly, and then decreases more slowly. This matches how the hours of daylight change over the course of a year.

Read and make a sketch for Situation C. The independent variable is the width of the garden, and the dependent variable is its area. Your sketch should be similar to Graph 1, which is a nonlinear graph that starts at 0, increases quickly at first, then slows down and peaks, and then decreases, slowly at first and then more quickly. When the width is 0, so is the area. This matches the description of how the area changes with increasing width.

Read and make a sketch for Situation D. The independent variable is the time, and the dependent variable is the difference between the tea temperature and the room

(continued)

temperature. Your sketch should be similar to Graph 4, which is a nonlinear graph that decreases quickly at first and then more and more slowly. This matches the description of the temperature change.

Now, think of another situation involving two related variables, and describe how one of the variables changes relative to the other. Sketch a graph to illustrate your situation, and write a description of your graph using the words *linear, nonlinear, increasing, decreasing, rate of change, peak,* and *minimum* or *least value.*

Functions that have smooth graphs, with no breaks in the domain or range, are called **continuous** functions. Functions that are not continuous often involve quantities—such as people, cars, or stories of a building—that are counted or measured in whole numbers. Such functions are called **discrete** functions. Read about continuous and discrete functions on page 455 of your book.

The example in your book shows how breaking the domain of a function into intervals can sometimes help you describe the function's behavior. Below is another example.

EXAMPLE Use the intervals marked on the x-axis in the graph below to help you discuss where the function is increasing or decreasing and where it is linear or nonlinear.

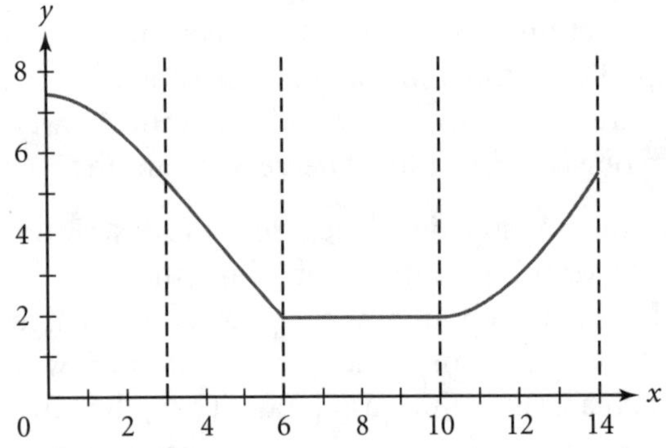

▶ ***Solution*** The function is decreasing in the interval $0 \leq x \leq 6$. In the interval $0 \leq x \leq 3$, the function is nonlinear and decreases slowly, then more quickly. In the interval $3 \leq x \leq 6$, the function is linear, decreasing at a constant rate.

In the interval $6 \leq x \leq 10$, the function is linear and is not increasing or decreasing.

In the interval $10 \leq x \leq 14$, the function is increasing and nonlinear. The function increases slowly at first and then more quickly.

CONDENSED LESSON

8.6 Defining the Absolute Value Function

In this lesson you will

- evaluate numerical expressions involving **absolute value**
- investigate the **absolute value function**
- solve equations involving the absolute value of a variable

The **absolute value** of a number is its size, or magnitude, regardless of whether the number is positive or negative. You can think of the absolute value of a number as its distance from zero on a number line. For example, both 7 and -7 are 7 units from zero, so both have an absolute value of 7.

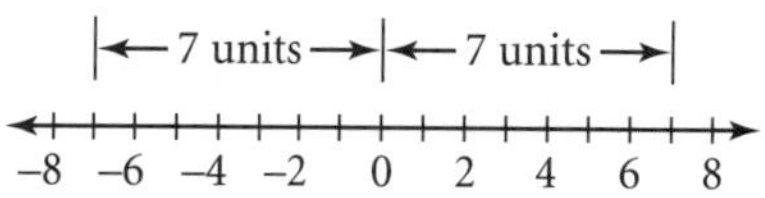

The notation $|x|$ is used to denote the absolute value of a number or an expression. So $|7| = 7$ and $|-7| = 7$. Read Example A in your book.

Investigation: Deviations from the Mean

In this investigation you will learn how the absolute value is useful for describing how much a data value deviates from the mean.

Students on the math team took a difficult test to try to qualify for a statewide competition. The first column of the table below shows their test scores. Enter the values from this column into list L_1 on your calculator.

The mean of the test scores is 29.8. The second column of the table shows the difference between each score and the mean. These values show how much each score *deviates* from the mean. Enter the formula $L_1 - 29.8$ into list L_2.

Here is a dot plot of the test scores.

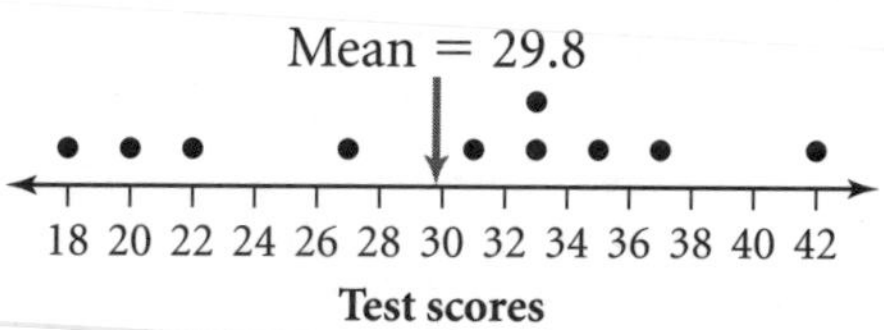

Score	Deviation (score − mean)	Distance from mean
27	−2.8	2.8
33	3.2	3.2
42	12.2	12.2
22	−7.8	7.8
37	7.2	7.2
20	−9.8	9.8
35	5.2	5.2
33	3.2	3.2
31	1.2	1.2
18	−11.8	11.8

The third column of the table shows the distance of each data value from the mean. These values are all positive. They are the *absolute values* of the deviations in the second column. Enter these values into list L_3.

Here is a scatter plot with the L_2 values (the deviations) on the x-axis and the L_3 values (the distances) on the y-axis. The x-values are both positive and negative, while the y-values are all positive. If you trace the graph, you will find that positive inputs are unchanged as outputs, while negative inputs are changed to their opposites.

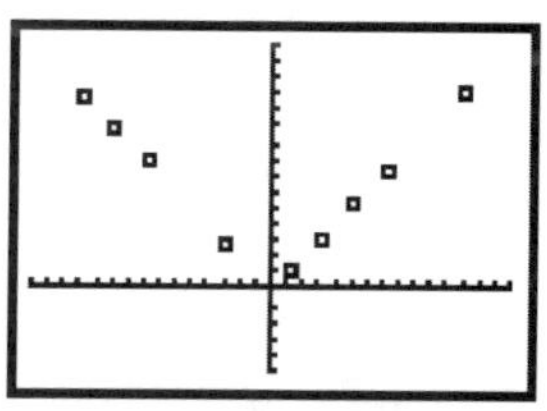

$[-15, 15, 1, -5, 15, 1]$

(continued)

The graph of $y = |x|$ (that is, $Y_1 = \text{abs}(x)$) has been added to the plot. Because each L_3 value *is* the absolute value of the corresponding L_2 value, this graph passes through all the points in the plot.

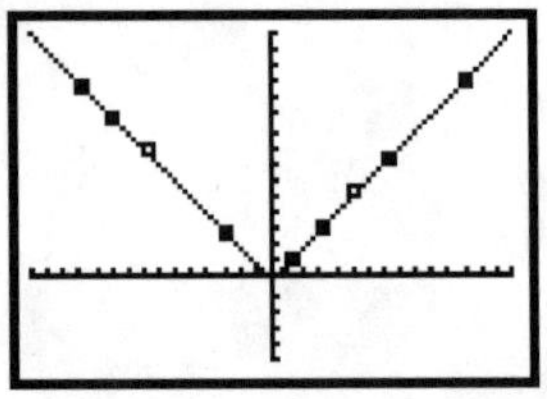

Find the mean of the deviations in list L_2 and of the distances in list L_3. The mean of the deviations is 0, and the mean of the distances is 6.44. When you are calculating the mean of the deviations, the positive and negative values "cancel each other out," resulting in a mean of 0. The mean of the distances, which are all positive, better indicates the variation in the data.

Write a rule for the absolute value function. Here is one possible rule: If x is positive or zero, then y is equal to x. If x is negative, then y is equal to the opposite of x.

Read the rest of the lesson in your book, paying close attention to Example B, which shows three methods for solving an equation involving the absolute value of a variable. Here is another example.

EXAMPLE | Solve the equation $-3|x| + 7 = -29$ symbolically.

▶ Solution

$-3\|x\| + 7 = -29$	Original equation.
$-3\|x\| = -36$	Subtract 7 from both sides.
$\|x\| = 12$	Divide both sides by -3.
$x = 12$ or $x = -12$	Find two numbers with absolute value 12.

CONDENSED LESSON

8.7 Squares, Squaring, and Parabolas

In this lesson you will

- calculate the **squares** of numbers
- investigate the **squaring function** and its graph
- use the **square root function** to undo the squaring function

When you multiply the number 4 by itself, you get 16. You get the same result if you multiply -4 by itself. The product of a number and itself is the **square** of the number, and the process of multiplying a number by itself is called **squaring.** The square of a number x is "x squared" and is written as x^2.

When you square numbers, you need to be careful of the order of operations. Calculate -5^2 and $(-5)^2$ on your calculator. For -5^2, your calculator squares 5 and then takes the opposite of the result, giving the output value -25. For $(-5)^2$, your calculator squares -5—that is, it multiplies -5 by itself—giving the output 25.

Investigation: Graphing a Parabola

Steps 1–5 Make a two-column table with the integers -10 through 10 in the first column and the square of each integer in the second column. Square the numbers *without* using your calculator. Here is a portion of the table.

Number (x)	Square (x^2)
-4	16
-3	9
-2	4
-1	1
0	0
1	1
2	4
3	9
4	16

Now, enter the integers -10 through 10 into list L1 on your calculator. Use the x^2 key to square each number. (See **Calculator Note 8B.**) Store the result in list L2. Make sure the values in list L2 match those in your table.

Look at the values in the table. Notice that the squares of both the positive and negative numbers are positive and that the square of a number is equal to the square of its opposite.

Make a scatter plot of the (L1, L2) values. Then, graph $y = x^2$ in the same window. The graph of $y = x^2$ shows the relationship between any number and its square.

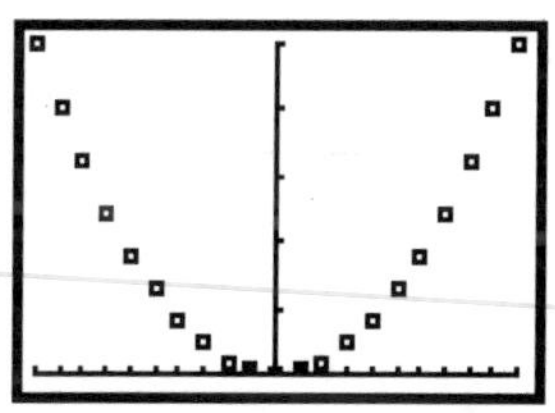

$[-12, 12, 1, -10, 110, 10]$

You can use the vertical line test to verify that $y = x^2$ is a function. The domain of this function is the set of all real numbers. The range is the set of real numbers greater than or equal to 0.

Steps 6–10 The graph of $y = x^2$ is a **parabola.** Except for (0, 0), all the points on the parabola are in Quadrants I and II. Looking at the graph, you can see that every output value, except 0, corresponds to two input values. For example, the output 25 corresponds to the inputs -5 and 5, and the output 6.25 corresponds to the inputs -2.5 and 2.5.

(continued)

A vertical line drawn through the origin divides the parabola into halves that are mirror images. This line (which is the y-axis) is called a *line of symmetry.* If you folded the graph along the line, the two halves of the parabola would match exactly.

Compare the parabola with the graph of $y = |x|$. Both graphs open upward, both are continuous, both have only positive y-values, and both have the y-axis as a line of symmetry. However, the parabola is curved at the bottom, whereas the absolute-value graph is pointed.

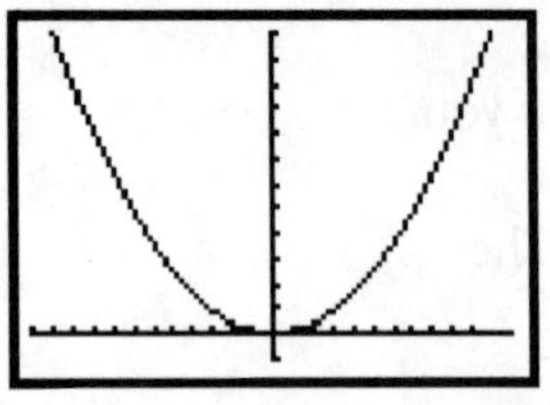

[−12, 12, 1, −10, 120, 10]

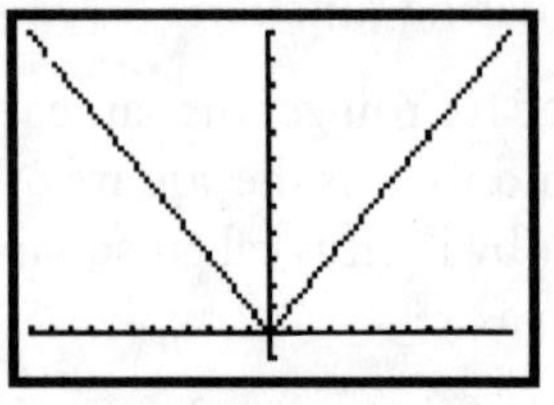

[−12, 12, 1, −1, 12, 1]

The x- and y-coordinates of each point on the parabola in the first quadrant could represent the side length and the area of a square. For example, the point (8, 64) represents a square with side length 8 and area 64.

Read the rest of the lesson in your book, which discusses how to "undo" the squaring function. Then, read the example below.

EXAMPLE Solve the equation $x^2 - 45 = 19$ symbolically.

► ***Solution***

$x^2 - 45 = 19$	Original equation.
$x^2 = 64$	Add 45 to both sides.
$\sqrt{x^2} = \sqrt{64}$	Take the square root of both sides.
$\lvert x \rvert = 8$	$\lvert x \rvert$ is the positive square root of $\sqrt{x^2}$, and 8 is the positive square root of 64.
$x = 8$ or $x = -8$	There are two solutions.

CONDENSED LESSON

9.1 Translating Points

In this lesson you will

- **translate** figures on the coordinate plane
- define a **translation** by describing how it affects a general point (x, y)

A mathematical rule that changes or moves a figure is called a **transformation.** In this lesson, you explore one type of transformation.

Investigation: Figures in Motion

Steps 1–6 The triangle below has vertices $(-1, 2)$, $(1, -1)$, and $(3, 1)$.

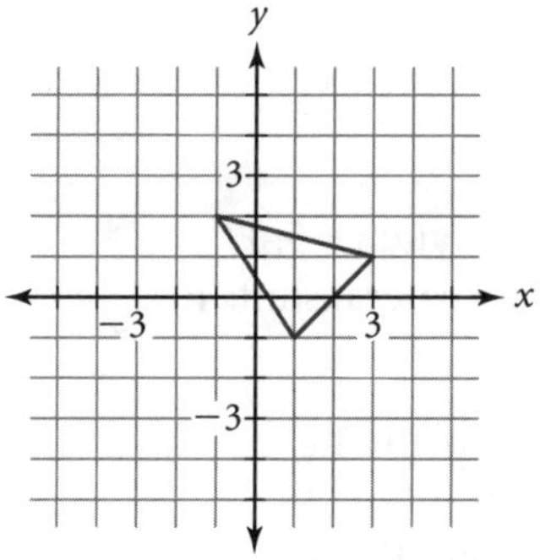

If you add 3 to each y-coordinate, you get $(-1, 5)$, $(1, 2)$, and $(3, 4)$. If you subtract 2 from each y-coordinate, you get $(-1, 0)$, $(1, -3)$, and $(3, -1)$. The grids below show the original triangle and triangles with vertices at the "transformed" points.

Add 3 to the y-coordinates

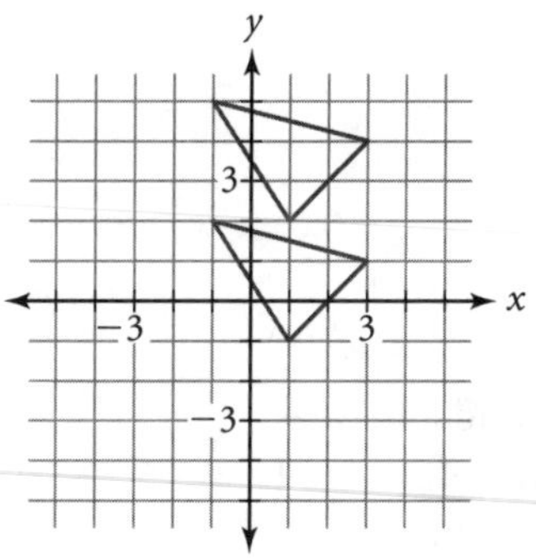

Subtract 2 from the y-coordinates

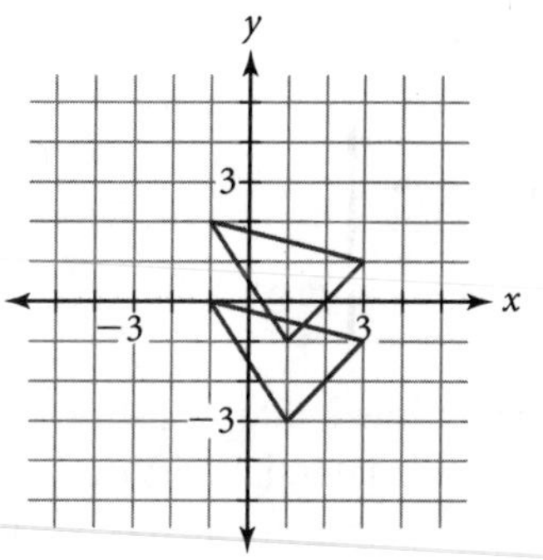

Notice that *adding* 3 to the y-coordinates moves the triangles *up* 3 units and that *subtracting* 2 from the y-coordinates moves the triangle *down* 2 units.

Now, draw your own triangle, and move it by adding or subtracting a number from the y-coordinates of the vertices.

Each grid in Step 6 on page 477 shows an original triangle and the triangle that results from adding a number to or subtracting a number from the y-coordinates. See if you can figure out what number was added or subtracted.

(continued)

Here are the answers to Step 6.

a. 3 was added. **b.** 4 was subtracted. **c.** 5 was subtracted.

Steps 7–13 Now, you will draw and move a polygon using your calculator. The vertices of the quadrilateral shown in Step 7 are $(1, 2)$, $(2, -2)$, $(-3, -1)$, and $(-2, 1)$. Follow Step 8 to enter the coordinates and draw the quadrilateral on your calculator.

Define lists L3 and L4 so that L3 = L1 − 3 and L4 = L2. So, L3 contains the original *x*-coordinates minus 3, and L4 contains the original *y*-coordinates. Graph a new quadrilateral using L3 for the *x*-coordinates and L4 for the *y*-coordinates.

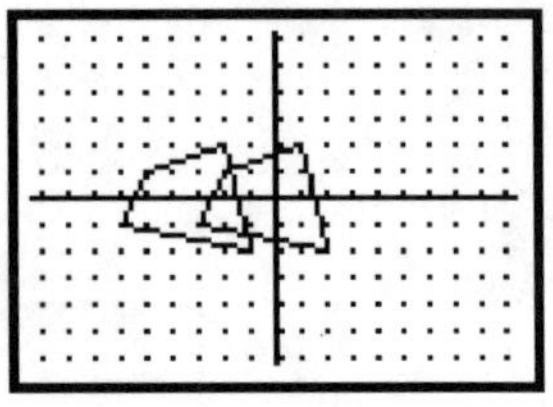

The vertices of the new quadrilateral are $(-2, 2)$, $(-1, -2)$, $(-6, -1)$, and $(-5, 1)$. Notice that subtracting 3 from the *x*-coordinates of the original quadrilateral shifts it 3 units to the left.

Follow Steps 9 and 10 at least two more times, adding a different number to or subtracting a different number from the *x*-coordinates each time. You should find that adding a positive number to the *x*-coordinates shifts the figure right that many units and that subtracting a positive number from the *x*-coordinates shifts the figure left that many units.

Now, let L3 = L1 − 1 and let L4 = L2 + 3. This subtracts 1 from the original *x*-coordinates and adds 3 to the original *y*-coordinates, shifting the quadrilateral left 1 unit and up 3 units as shown below.

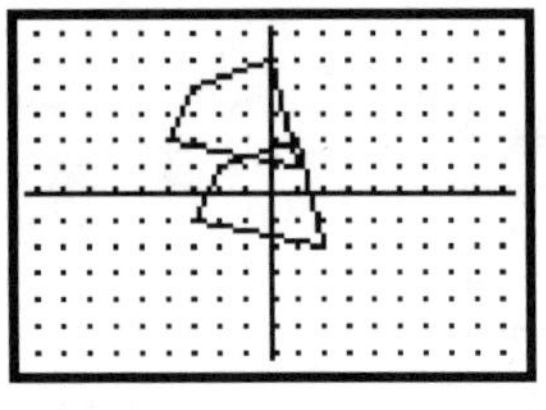

Each graphing window in Step 12 shows the original quadrilateral and a new, "transformed" quadrilateral. Write definitions for L3 and L4 in terms of L1 and L2 that would create the transformed quadrilateral. Here are the correct rules.

a. L3 = L1 + 6, L4 = L2 **b.** L3 = L1 − 3, L4 = L2 **c.** L3 = L1 − 5, L4 = L2 + 3

When you transform a figure, the result is called the **image** of the original figure. Horizontal and vertical transformations, like the ones you explored in the investigation, are called **translations.** You can define a translation by describing the image of a general point (x, y). For example, the translation that shifts a figure left 4 units and up 2 units can be defined as $(x - 4, y + 2)$.

Now, read and follow along with the example in your book.

CONDENSED LESSON

9.2 Translating Graphs

In this lesson you will

- **translate** the **absolute value** and **squaring functions**
- **translate** an **exponential function**
- learn about **families of functions**

In the previous lesson you translated figures on the coordinate plane. In this lesson you will learn how to translate functions.

Investigation: Translations of Functions

Steps 1–6 If you substitute $x - 3$ for x in the absolute value function $y = |x|$, you get $y = |x - 3|$. You can think of this as finding $f(x - 3)$ when $f(x) = |x|$. Enter $y = |x|$ into Y1 and $y = |x - 3|$ into Y2, and graph both functions.

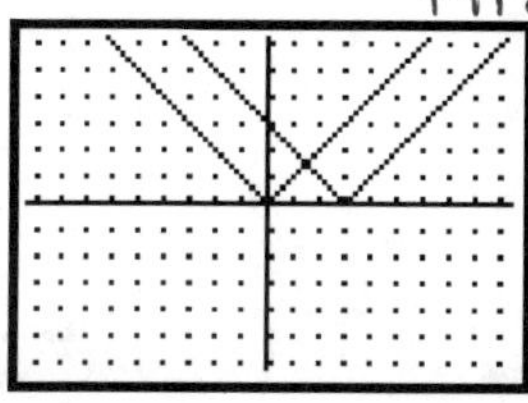

Notice that the graph of $y = |x - 3|$ is the graph of $y = |x|$ translated right 3 units.

The **vertex** of an absolute value graph is the point where the function changes from decreasing to increasing or from increasing to decreasing. The vertex of $y = |x|$ is (0, 0), and the vertex of $y = |x - 3|$ is (3, 0). So, the vertex of $y = |x|$, like the rest of the graph, has been translated right 3 units.

The function $y = |x - (-4)|$ or $y = |x + 4|$ translates the graph of $y = |x|$ left 4 units. To get $y = |x + 4|$, you substitute $x + 4$ for x in $y = |x|$.

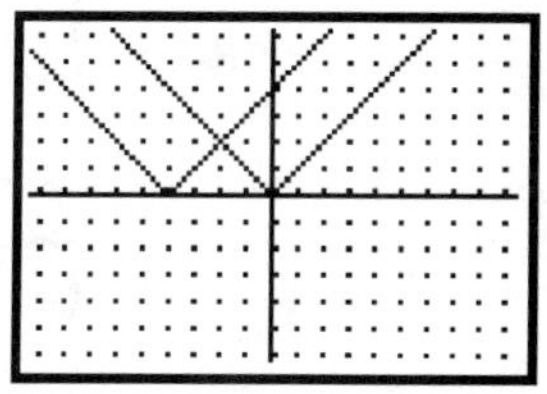

Write a function to create each translation of $y = |x|$ shown in Step 6. Use your calculator to check your work. You should get these results.

a. $y = |x - 2|$ **b.** $y = |x - 5|$ **c.** $y = |x + 3|$

Steps 7–12 Now, you will translate $y = |x|$ along the y-axis.

If you substitute $y - 3$ for y in $y = |x|$, you get $y - 3 = |x|$ or (solving for y) $y = 3 + |x|$. Here are the graphs of $y = |x|$ and $y = 3 + |x|$ on the same axes.

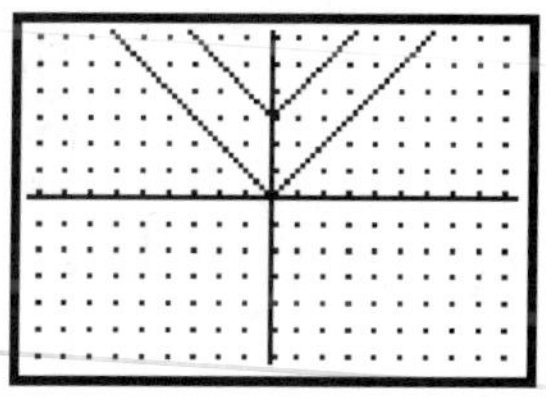

Notice that the graph of $y = 3 + |x|$ is the graph of $y = |x|$ translated up 3 units. The vertex of $y = 3 + |x|$ is (0, 3), which is the vertex of $y = |x|$ translated up 3 units.

If you substitute $y - (-3)$ or $y + 3$ for y in $y = |x|$, you get $y + 3 = |x|$ or $y = -3 + |x|$. The graph of $y = -3 + |x|$ is the graph of $y = |x|$ translated down 3 units.

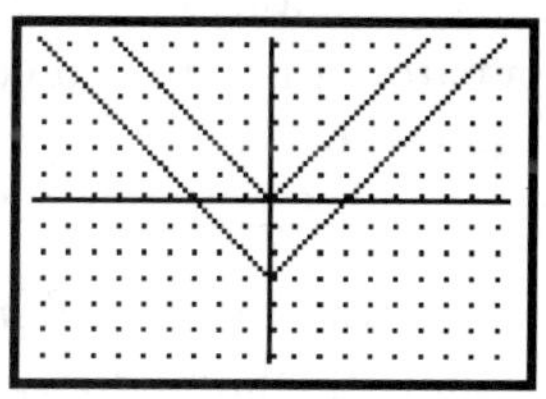

Write a function to create each translation of $y = |x|$ shown in Step 12. Use your calculator to check your work. You should get these results.

a. $y = -2 + |x|$ **b.** $y = 1 + |x|$ **c.** $y = -4 + |x - 3|$

(continued)

Step 13 You have seen that to translate the graph of $y = |x|$ horizontally, you subtract a number from x in the function. Subtracting a positive number translates the graph right, and subtracting a negative number translates the graph left. To translate the graph vertically, you add a number to the entire function. Adding a positive number translates the graph up, and adding a negative number translates the graph down.

EXAMPLE

The same ideas you used to translate the absolute value function can be used to translate the function $y = x^2$.

Here is the graph of $y = x^2$ and $y = (x + 2)^2$.

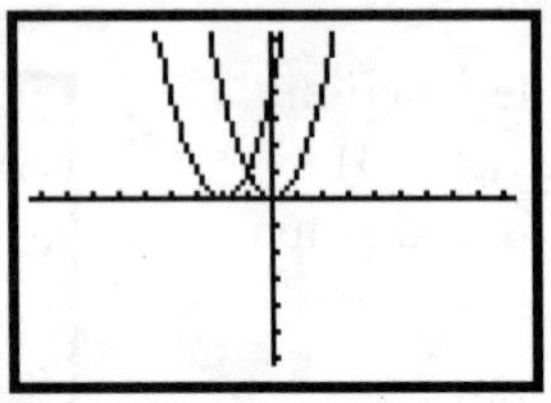

Here is the graph of $y = x^2$ and $y = x^2 - 3$.

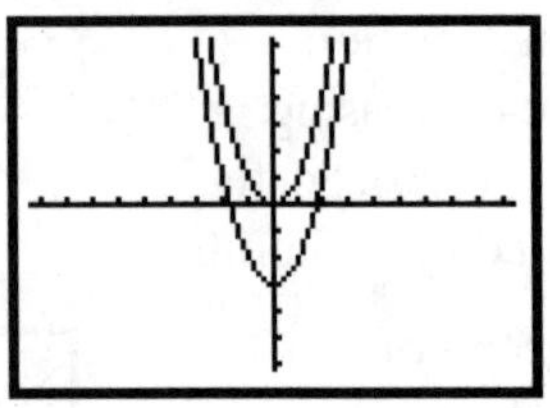

Here is the graph of $y = x^2$ and $y = (x + 5)^2 + 2$.

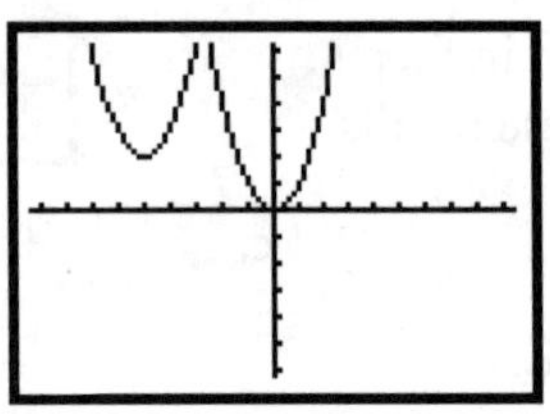

The vertex of a parabola is the point where the graph changes from decreasing to increasing or from increasing to decreasing. The vertices of the parabolas above are $(-2, 0)$, $(0, 3)$, and $(-5, 2)$. Notice that the x-coordinate of the vertex is the value subtracted from x in the function and that the y-coordinate is the value added to the entire function.

The functions $y = |x|$ and $y = x^2$ are examples of **parent functions.** By transforming a parent function, you can create infinitely many functions in the same **family of functions.** For example, the functions $y = (x + 2)^2$ and $y = x^2 - 4$ are both part of the squaring family of functions, which has $y = x^2$ as the parent function.

Now, read Example B in your book, which shows you how to translate an exponential function.

CONDENSED LESSON

9.3 Reflecting Points and Graphs

In this lesson you will

- **reflect polygons** over the x- and y-axes
- **reflect graphs of functions** over the x- and y-axes
- write equations for graphs created by **combining transformations**

You have studied translations—transformations that slide a figure horizontally or vertically. In this lesson you will learn about transformations that flip a figure across a line.

Investigation: Flipping Graphs

Steps 1–5 The triangle on page 492 of your book has vertices (1, 5), (3, 1), and (6, 2). The grid below shows the original triangle and the triangle formed by changing the sign of the x-coordinate of each vertex to get (−1, 5), (−3, 1), and (−6, 2).

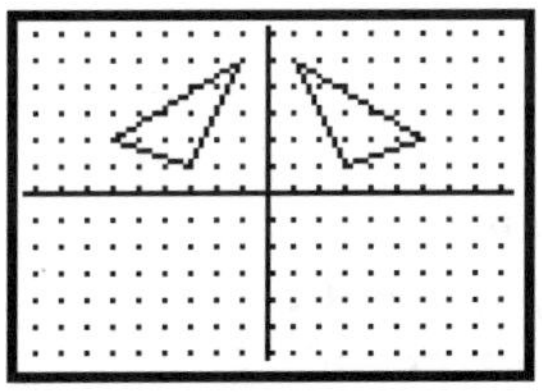

Changing the signs of the x-coordinates *flips* a figure across the y-axis, creating a mirror image of the original. You can match the original figure exactly with the image by folding the grid along the y-axis.

The grid below shows the original triangle and the triangle formed by changing the sign of the y-coordinate of each vertex to get (1, −5), (3, −1), and (6, −2).

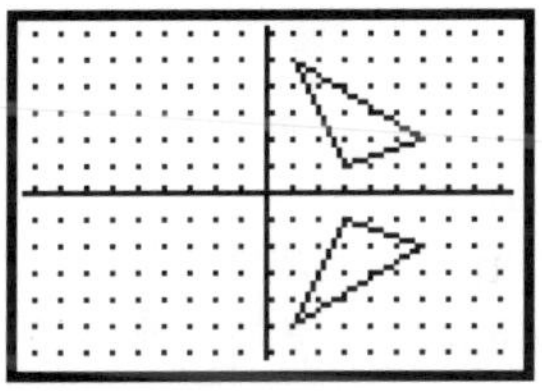

Changing the signs of the y-coordinates *flips* a figure across the x-axis, creating a mirror image of the original. You can match the original figure exactly with the image by folding the grid along the x-axis.

You change the signs of both the x- and y-coordinates to get (−1, −5), (−3, −1), and (−6, −2). This transformation flips the figure across one axis and then across the other.

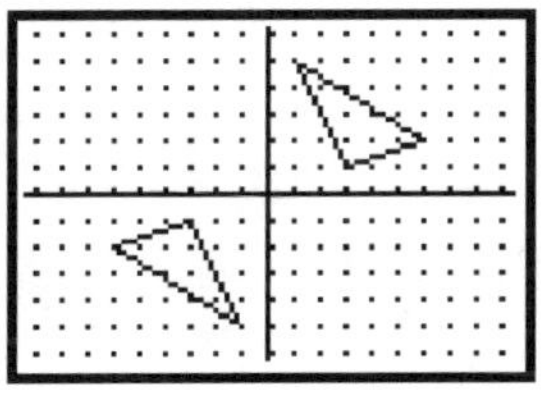

(continued)

Steps 6–10 If you replace the x in the function $y = 2^x$ with $-x$, you get $y = 2^{-x}$. (This is the same as finding $f(-x)$ when $f(x) = 2^x$.) If you graph both functions on your calculator, you'll see that the graph of $y = 2^{-x}$ is the graph of $y = 2^x$ flipped across the y-axis.

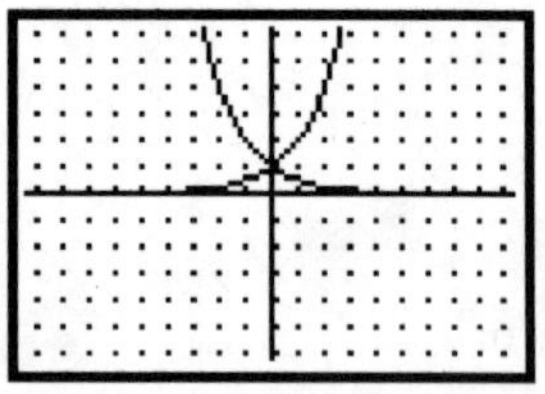

If you replace the y in $y = 2^x$ with $-y$, you get $-y = 2^x$ or (solving for y) $y = -2^x$. [This is the same as finding $-f(x)$ when $f(x) = 2^x$.] If you graph both functions on your calculator, you'll see that the graph of $y = -2^x$ is the graph of $y = 2^x$ flipped across the x-axis.

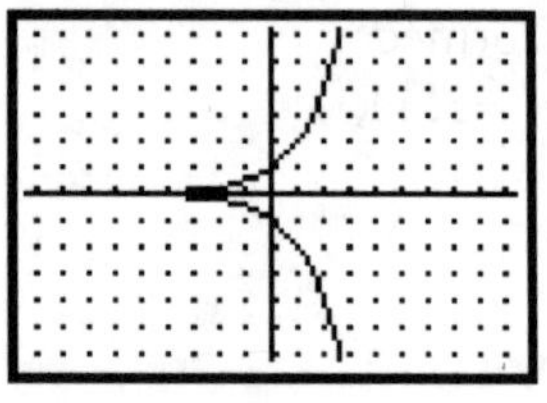

For each function in Step 9, replace x with $-x$, and graph both the original function and the new function. In each case you should find that the graph of the original function is flipped across the y-axis to get the graph of the new function. (Note: Because the graph of $y = |x|$ is symmetric across the y-axis, it looks the same when it is flipped across the y-axis. So, the graphs of $y = |x|$ and $y = |-x|$ are identical.)

For each function in Step 9, replace y with $-y$, and graph both the original function and the new function. In each case you should find that the graph of the original function is flipped across the x-axis to get the graph of the new function.

A transformation that flips a figure to create a mirror image is called a **reflection.** As you discovered in the investigation, a point is **reflected across the x-axis** when you change the sign of its y-coordinate. A point is **reflected across the y-axis** when you change the sign of its x-coordinate. Similarly, a function is reflected across the x-axis when you change the sign of y, and a function is reflected across the y-axis when you change the sign of x.

Read the rest of the lesson in your book. Read the example very carefully. It explains how to write equations for graphs created by applying more than one transformation to the graph of a parent function.

CONDENSED LESSON

9.4 Stretching and Shrinking Graphs

In this lesson you will

- **stretch** and **shrink** a **quadrilateral**
- **stretch** and **shrink** the graph of a **function**
- write equations for graphs formed by **combining transformations**

You have learned about transformations that slide a figure horizontally or vertically and that flip a figure across a line. In this lesson you'll study a transformation that changes a figure's shape.

Investigation: Changing the Shape of a Graph

Steps 1–5 Copy the quadrilateral on page 502 of your book onto graph paper or enter the x-coordinates into list L1 and the y-coordinates into list L2. The coordinates of the vertices are $(1, 3)$, $(2, -1)$, $(-3, 0)$, and $(-2, 2)$.

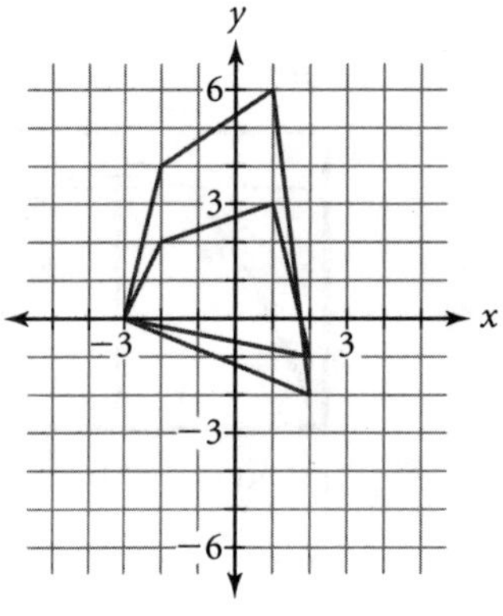

Multiply the y-coordinate of each vertex by 2 to get $(1, 6)$, $(2, -2)$, $(-3, 0)$, and $(-2, 4)$. On the same grid as the original figure, draw a new quadrilateral with these new points as vertices. Or follow the instructions in your book to graph it on your calculator.

As you can see, multiplying the y-coordinates by 2 stretches the figure vertically. Points above the x-axis move up. Points below the x-axis move down. Points on the x-axis remain fixed.

Now, multiply the y-coordinates of the vertices of the original quadrilateral by 3, by 0.5, and by -2, and draw the resulting quadrilaterals. Below are the results for 0.5 and -2.

Multiply by 0.5

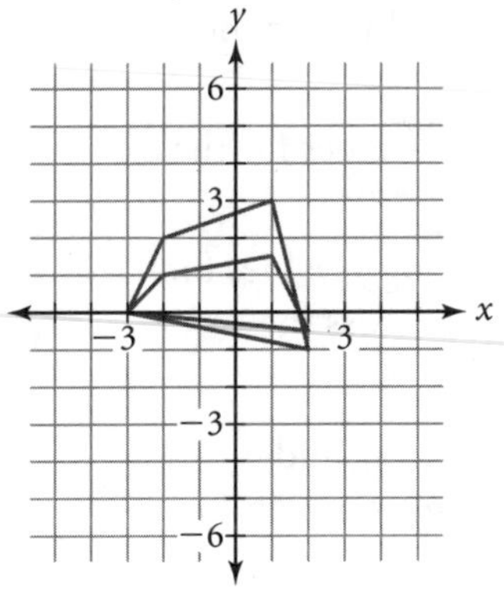

Multiply by -2

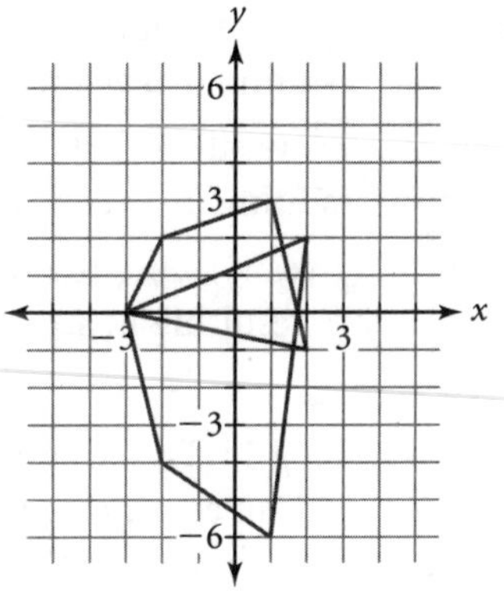

Multiplying by 0.5 shrinks the figure vertically to half its size. Multiplying by -2 stretches the figure vertically and flips it over the x-axis. In general, multiplying the y-coordinates of a figure by a number a

- stretches the figure if $|a| > 1$ and shrinks the figure if $|a| < 1$
- reflects the figure across the x-axis if $a < 0$

(continued)

Steps 6–8 Graph the triangle shown in Step 6 on your calculator, putting the x-coordinates in list L_1 and the y-coordinates in list L_2. Then, predict how the definitions in parts a and b of Step 7 will transform the triangle, and use your calculator to check your answers. Here are the results.

a. The figure shrinks vertically to half its size and is flipped over the x-axis.

b. The figure stretches vertically to twice its size and is translated down 2 units.

In Step 8, write definitions for lists L_3 and L_4 that would create each image. Here are the answers.

a. $L_3 = L_1$; $L_4 = 3 \cdot L_2$ **b.** $L_3 = L_1$; $L_4 = 2 \cdot L_2 + 3$

Steps 9–13 Enter the equation $f(x) = -x^2 + 1$ as Y_1, and graph it on your calculator. Multiply the right side of the equation by 2—that is, find $y = 2 \cdot f(x)$—and enter the result, $y = 2(-x^2 + 1)$, as Y_2. Below are the graph and table for the two functions.

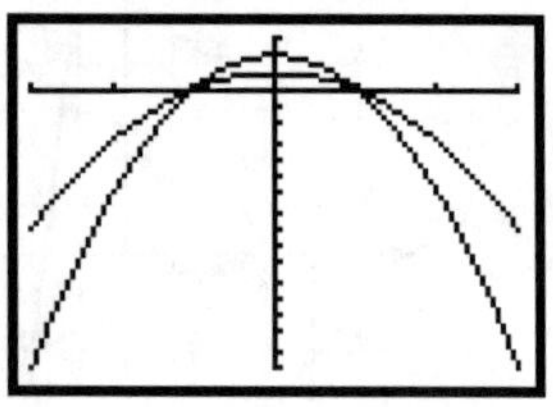

X	Y1	Y2
-3	-8	-16
-2	-3	-6
-1	0	0
0	1	2
1	0	0
2	-3	-6
3	-8	-16

X= -3

[−3, 3, 1, −16, 3, 1]

In the table, notice that each y-value for $y = 2(-x^2 + 1)$ is twice the corresponding y-value for $y = 1 - x^2$. This causes the graph to stretch. The points on the graph of $y = 2(-x^2 + 1)$ are twice as far from the x-axis as the corresponding points on the graph of $y = -x^2 + 1$. Multiplying by 2 causes the points above the x-axis to move up and the points below the x-axis to move down.

Repeat the process above for $y = 0.5 \cdot f(x)$, $y = 3 \cdot f(x)$, and $y = -2 \cdot f(x)$. You should find the following:

- Multiplying by 0.5 gives y-values that are half the corresponding y-values for the original function, resulting in a vertical shrink.
- Multiplying by 3 gives y-values that are 3 times the corresponding y-values for the original function, resulting in a vertical stretch.
- Multiplying by -2 gives y-values that are -2 times the corresponding y-values for the original function, resulting in a vertical stretch and a reflection across the x-axis.

Look at the graphs in Step 13. Use what you have learned about stretching and shrinking graphs to write an equation for $R(x)$ in terms of $B(x)$ and an equation for $B(x)$ in terms of $R(x)$. You should get the following results:

a. $R(x) = 3 \cdot B(x)$; $B(x) = \frac{1}{3} \cdot R(x)$

b. $R(x) = -\frac{1}{2} \cdot B(x)$; $B(x) = -2 \cdot R(x)$

Read the rest of the lesson, including the examples, very carefully.

CONDENSED LESSON

9.6 Introduction to Rational Functions

In this lesson you will

- explore transformations of the parent function $y = \frac{1}{x}$
- predict what the graph of a **rational function** looks like based on its equation

In Chapter 3, you learned about inverse variation. The simplest inverse variation equation is $y = \frac{1}{x}$. On page 513 of your book, read about $y = \frac{1}{x}$ and its graph. In this lesson you will see how the parent function $y = \frac{1}{x}$ can help you understand many other functions.

Investigation: I'm Trying to Be Rational

Steps 1–5 Graph the parent function $y = \frac{1}{x}$ on your calculator. The functions in Step 2 are all transformations of $y = \frac{1}{x}$. Use what you have learned in this chapter to predict how the graph of each function will compare to the graph of $y = \frac{1}{x}$. The answers are given below.

a. Vertical stretch by a factor of 3 and reflection across the x-axis

b. Vertical stretch by a factor of 2 and translation up 3 units

c. Translation right 2 units

d. Translation left 1 unit and down 2 units

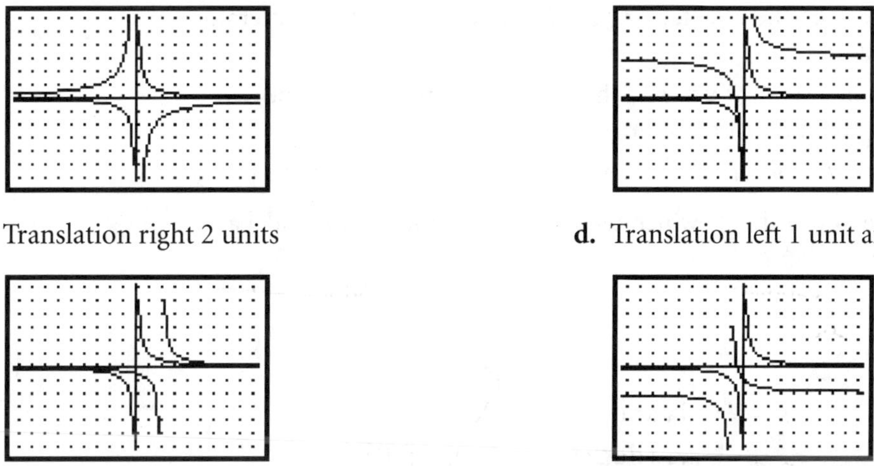

Now, without graphing, describe what the graph of each function in Step 4 looks like. Use the words *linear, nonlinear, increasing,* and *decreasing.* Define the domain and range, and give equations for the asymptotes. Here are sample answers.

Graph a is a vertical stretch of $y = \frac{1}{x}$ by a factor of 5, followed by a translation right 4 units; it is nonlinear and decreasing; the domain is $x \neq 4$, and the range is $y \neq 0$; there are asymptotes at $x = 4$ and $y = 0$.

Graph b is a reflection of $y = \frac{1}{x}$ across the x-axis, followed by a translation left 3 units and down 5 units; it is nonlinear and increasing; the domain is $x \neq -3$, and the range is $y \neq -5$; there are asymptotes at $x = -3$ and $y = -5$.

Graph c is a vertical stretch of $y = \frac{1}{x}$ by a factor of $|a|$ (if $a < 0$, the graph is also reflected across the x-axis), followed by a translation h units horizontally and k units vertically; it is nonlinear and decreasing if a is positive and nonlinear and increasing if a is negative; the domain is $x \neq h$, and the range is $y \neq k$; there are asymptotes at $x = h$ and $y = k$.

(continued)

Functions such as $y = \frac{5}{x-4}$ are called **rational functions** because they involve ratios of two expressions. Not all rational functions are transformations of $y = \frac{1}{x}$, but the graph of any rational function shares some similarities with the graph of $y = \frac{1}{x}$.

Steps 6–12 Graph $y = \frac{x+3}{(x-2)(x+3)}$ on your calculator. Compare the graph to the graphs of the functions in Step 2. You should find that the graph looks like the graph of $y = \frac{1}{x-2}$.

If you trace the graph, you should find that there is a "hole" at $x = -3$. The hole occurs because the function is not defined at $x = -3$ (the denominator is 0 at this value). However, for $x \neq -3$,

$$y = \frac{x+3}{(x-2)(x+3)} = \frac{1}{x-2} \cdot \frac{x+3}{x+3} = \frac{1}{x-2} \cdot 1 = \frac{1}{x-2}$$

So, $y = \frac{x+3}{(x-2)(x+3)}$ is identical to $y = \frac{1}{x-2}$ for $x \neq -3$ and is undefined for $x = -3$.

Now, graph the equation in Step 9. You should find that it has vertical asymptotes at $x = -5$ and $x = 1$ and a horizontal asymptote at $y = 4$.

In general, the following is true about the graph of a rational function:

- A hole occurs at $x = h$, when the factor $(x - h)$ occurs in *both* the numerator and the denominator.
- A vertical asymptote occurs at $x = h$, if the factor $(x - h)$ occurs in the denominator but *not* in the numerator.
- A horizontal asymptote occurs at $y = k$, when k is the constant of the equation.

Now, read the example in your book, which uses a rational function to model a real-world situation.

CONDENSED LESSON

9.7 Transformations with Matrices

In this lesson you will

- relate translations to **matrix addition**
- relate reflections, stretches, and shrinks to **matrix multiplication**

In your book, read the text before the investigation, which explains how to use a matrix to represent the vertices of a geometric figure.

Investigation: Matrix Transformations

Steps 1–6 Matrix $[A] = \begin{bmatrix} -4 & 3 & 2 \\ -1 & 4 & 0 \end{bmatrix}$ represents a triangle with vertices $(-4, -1)$, $(3, 4)$, and $(2, 0)$.

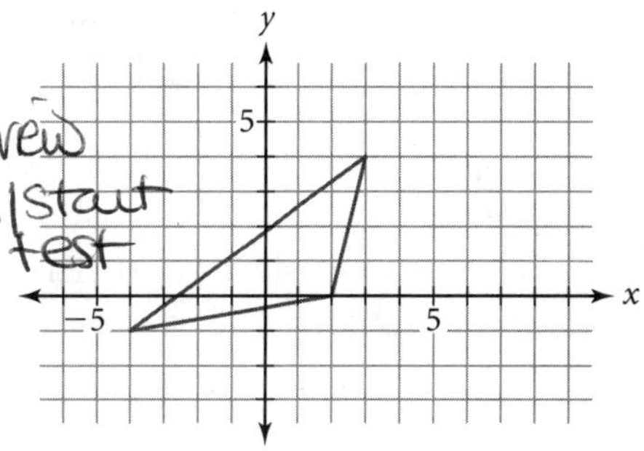

Adding the matrix $\begin{bmatrix} 5 & 5 & 5 \\ 0 & 0 & 0 \end{bmatrix}$ to matrix $[A]$ adds 5 to each x-coordinate, which translates the triangle right 5 units:

$$[A] + \begin{bmatrix} 5 & 5 & 5 \\ 0 & 0 & 0 \end{bmatrix} = \begin{bmatrix} -4 & 3 & 2 \\ -1 & 4 & 0 \end{bmatrix} + \begin{bmatrix} 5 & 5 & 5 \\ 0 & 0 & 0 \end{bmatrix} = \begin{bmatrix} 1 & 8 & 7 \\ -1 & 4 & 0 \end{bmatrix}$$

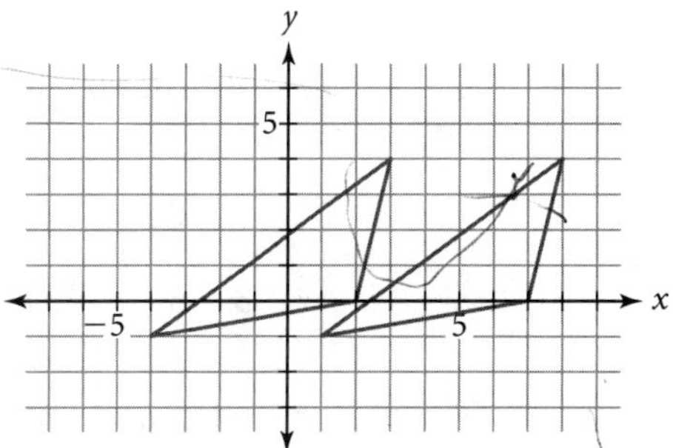

In parts a–c of Step 5, find the matrix sum and graph the resulting triangle. You should find that the sums correspond to the following translations of the original triangle:

a. down 4 units **b.** right 5 units, down 4 units **c.** left 6 units, up 4 units

Now, write matrix equations to represent the translations in Step 6. Here are the answers.

a. $\begin{bmatrix} 0 & 0 & 1 & 1 \\ 0 & 1 & 1 & 0 \end{bmatrix} + \begin{bmatrix} 2 & 2 & 2 & 2 \\ 1.5 & 1.5 & 1.5 & 1.5 \end{bmatrix} = \begin{bmatrix} 2 & 2 & 3 & 3 \\ 1.5 & 2.5 & 2.5 & 1.5 \end{bmatrix}$

b. $\begin{bmatrix} -3 & -2 & 1 & 2 \\ -1 & 1 & 2 & -2 \end{bmatrix} + \begin{bmatrix} -10.4 & -10.4 & -10.4 & -10.4 \\ 6.9 & 6.9 & 6.9 & 6.9 \end{bmatrix} = \begin{bmatrix} -13.9 & -12.4 & -9.4 & -8.4 \\ 5.9 & 7.9 & 8.9 & 4.9 \end{bmatrix}$

(continued)

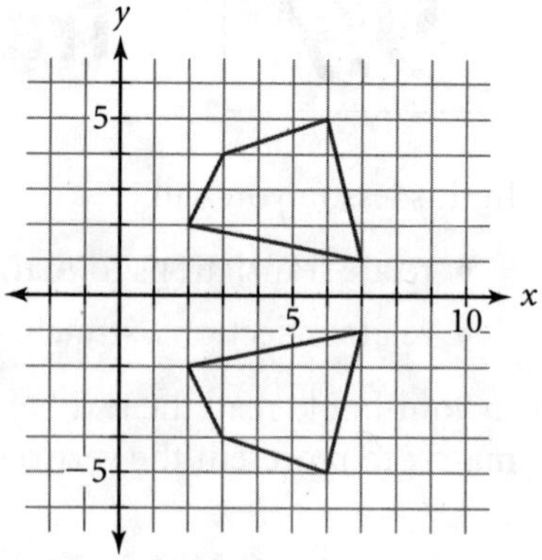

Steps 7–11 Copy the quadrilateral on page 521 of your book onto graph paper. Matrix $[B] = \begin{bmatrix} 2 & 3 & 6 & 7 & x \\ 2 & 4 & 5 & 1 & y \end{bmatrix}$ gives the coordinates of the vertices, along with the coordinates of a general point, (x, y).

Calculating $\begin{bmatrix} 1 & 0 \\ 0 & -1 \end{bmatrix} \cdot [B]$ multiplies each y-coordinate by -1, which reflects the quadrilateral across the x-axis:

$$\begin{bmatrix} 1 & 0 \\ 0 & -1 \end{bmatrix}\begin{bmatrix} 2 & 3 & 6 & 7 & x \\ 2 & 4 & 5 & 1 & y \end{bmatrix} = \begin{bmatrix} 2 & 3 & 6 & 7 & x \\ -2 & -4 & -5 & -1 & -y \end{bmatrix}$$

Now, perform the multiplications in Step 11 and graph the resulting quadrilaterals. You should get the following results:

a. $\begin{bmatrix} -1 & 0 \\ 0 & 1 \end{bmatrix}\begin{bmatrix} 2 & 3 & 6 & 7 & x \\ 2 & 4 & 5 & 1 & y \end{bmatrix} = \begin{bmatrix} -2 & -3 & -6 & -7 & -x \\ 2 & 4 & 5 & 1 & y \end{bmatrix}$. The x-coordinates are multiplied by -1. This reflects the quadrilateral across the y-axis.

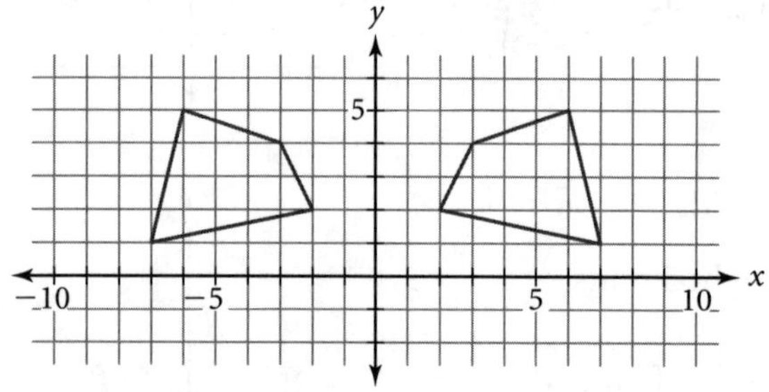

b. $\begin{bmatrix} 1 & 0 \\ 0 & 0.5 \end{bmatrix}\begin{bmatrix} 2 & 3 & 6 & 7 & x \\ 2 & 4 & 5 & 1 & y \end{bmatrix} = \begin{bmatrix} 2 & 3 & 6 & 7 & x \\ 1 & 2 & 2.5 & 0.5 & 0.5y \end{bmatrix}$.

The y-coordinates are multiplied by 0.5. This shrinks the quadrilateral vertically by a factor of 0.5.

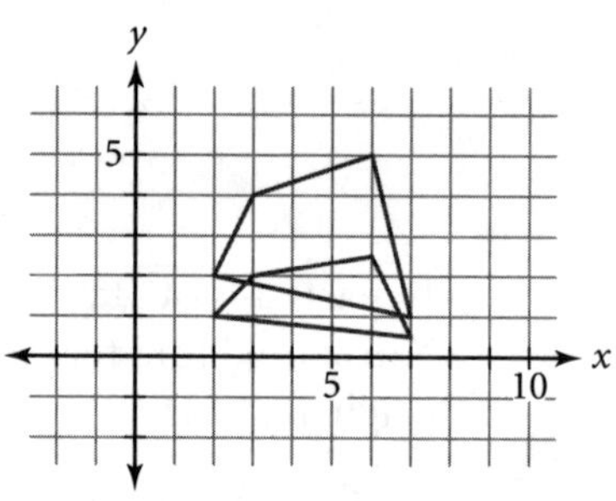

c. $\begin{bmatrix} 0.5 & 0 \\ 0 & 2 \end{bmatrix}\begin{bmatrix} 2 & 3 & 6 & 7 & x \\ 2 & 4 & 5 & 1 & y \end{bmatrix} = \begin{bmatrix} 1 & 1.5 & 3 & 3.5 & x \\ 4 & 8 & 10 & 2 & 2y \end{bmatrix}$.

The x-coordinates are multiplied by 0.5, and the y-coordinates are multiplied by 2. This shrinks the quadrilateral horizontally by a factor of 0.5 and stretches it vertically by a factor of 2.

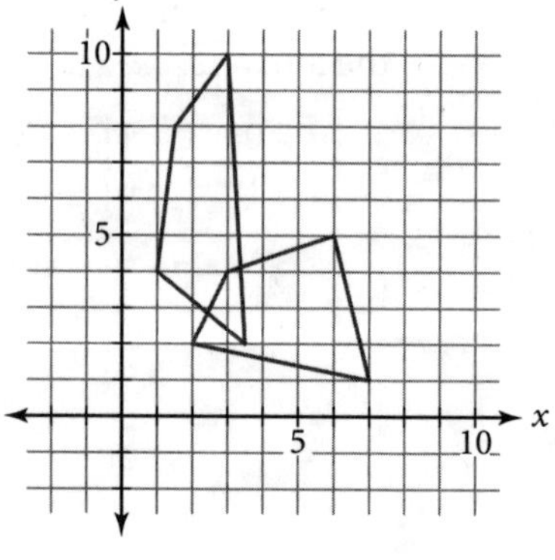

CONDENSED LESSON

10.1 Solving Quadratic Equations

In this lesson you will

- look at **quadratic functions** that model **projectile motion**
- use tables and graphs to approximate solutions to **quadratic equations**
- **solve quadratic equations** by undoing the order of operations

When an object is projected straight up into the air, its position at any time depends on its starting height, its initial velocity, and the force of gravity. If you plot the object's height at each instant of time, the resulting graph is a parabola. Read Example A in your book and look at the graph of the height of a popped-up baseball over time.

The motion of an object projected into the air can be modeled by a **quadratic function.** A quadratic function is any transformation of the parent function, $f(x) = x^2$.

Investigation: Rocket Science

A model rocket blasts off from a position 2.5 meters above the ground with a starting velocity of 49 m/sec. If the rocket travels straight up, and if the only force acting on it is gravity, then the **projectile motion** of the rocket can be described by the function

$$h(t) = \frac{1}{2}(-9.8)t^2 + 49t + 2.5$$

where t is the number of seconds after blastoff, and $h(t)$ is the height at time t.

The fact that $h(0) = 2.5$ means that the initial height of the rocket is 2.5 meters.

In metric units, the acceleration due to gravity is 9.8 m/sec^2. This value appears in the t^2 term of the equation $\frac{1}{2}(-9.8)t^2$. The negative symbol shows that the force is downward.

Graph the function on your calculator. Be sure to use a window that shows all the important features of the parabola. Here is the graph in the window $[-1, 12, 1, -10, 150, 10]$.

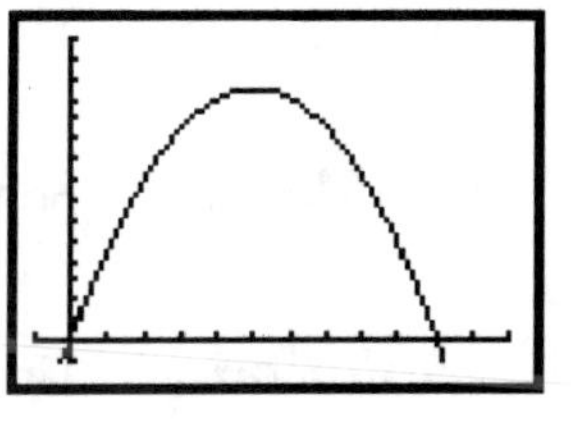

You can trace the graph to find the coordinates of the highest point (the vertex).

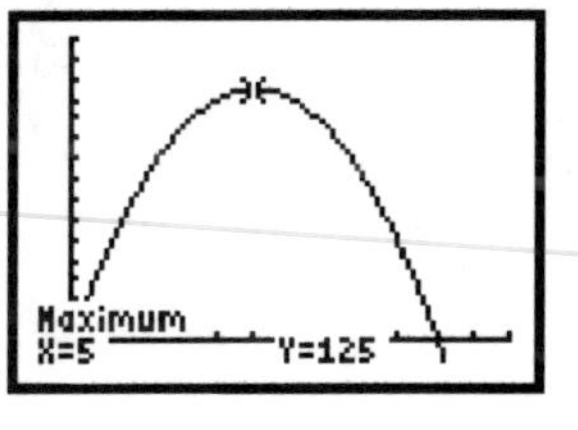

The coordinates are (5, 125), indicating that the rocket reaches a maximum height of 125 meters 5 seconds after blastoff.

To find the amount of time the rocket is in flight, find the coordinates of the point where the graph intersects the positive x-axis. The coordinates are (10.05, 0), indicating that the rocket hits the ground (that is, the height is 0) after 10.05 seconds. So, the flight of the rocket lasts just over 10 seconds.

To find the value of t when $h(t) = 50$, you would need to solve

$$\frac{1}{2}(-9.8)t^2 + 49t + 2.5 = 50$$

(continued)

A calculator table shows that the approximate solutions to this equation are 1.1 and 8.9, indicating that the rocket is at height 50 meters after about 1.1 seconds (when it is on its way up) and after about 8.9 seconds (when it is on its way down).

X	Y1	
1.04	48.16	
1.05	48.548	
1.06	48.934	
1.07	49.32	
1.08	49.705	
1.09	50.088	
1.1	50.471	
X=1.09		

X	Y1	Y2
8.87	51.613	50
8.88	51.233	50
8.89	50.853	50
8.9	50.471	50
8.91	50.088	50
8.92	49.705	50
8.93	49.32	50
X=8.91		

On a graph, the solutions are the x-coordinates of the points where the graphs of $h(t) = \frac{1}{2}(-9.8)t^2 + 49t + 2.5$ and $h(t) = 50$ intersect.

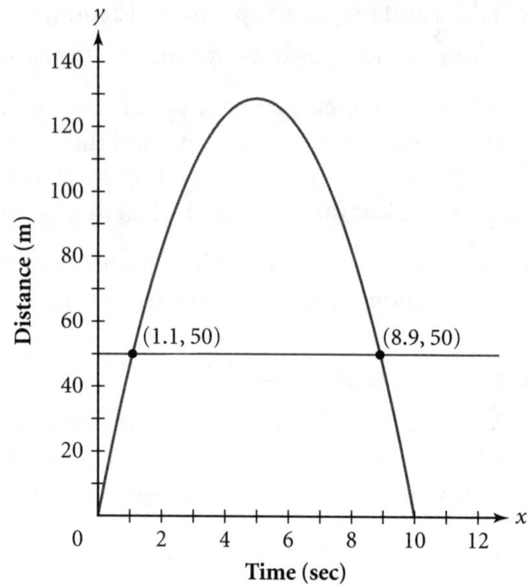

In the investigation, you approximated solutions to a quadratic equation using tables and graphs. To solve a quadratic equation using the symbolic methods you are familiar with, you must put the equation in a particular form. Example B in your book solves a quadratic equation by "undoing" the order of operations. Below is another example. (Note: Later, you will learn new methods that let you solve any quadratic equation.)

EXAMPLE | Solve $-2(x - 1)^2 + 9 = 4$ symbolically.

▶ ***Solution***

Undo each operation as you would when solving a linear equation. Keep the equation balanced by doing the same thing to both sides. To undo squaring, take the square root of both sides.

$-2(x-1)^2 + 9 = 4$	The original equation.
$-2(x-1)^2 = -5$	Subtract 9 from both sides.
$(x-1)^2 = 2.5$	Divide both sides by -2.
$\sqrt{(x-1)^2} = \sqrt{2.5}$	Take the square root of both sides.
$x - 1 = \pm\sqrt{2.5}$	The $\pm$ symbol shows the two numbers $+\sqrt{2.5}$ and $-\sqrt{2.5}$ whose square is 2.5.
$x = 1 \pm \sqrt{2.5}$	Add 1 to both sides.

The two solutions are $1 + \sqrt{2.5}$ and $1 - \sqrt{2.5}$, or approximately 2.58 and -0.58.

Read the rest of the investigation in your book.

CONDENSED
LESSON

10.2 Finding the Roots and the Vertex

In this lesson you will

- model a real-word situation with a quadratic function
- identify the ***x*-intercepts, vertex,** and **line of symmetry** of a parabola
- rewrite a quadratic function in **vertex form**

You have looked at quadratic functions that model projectile motion. In this lesson you'll explore another situation that can be modeled with a quadratic function.

Investigation: Making the Most of It

Suppose you have 24 meters of fencing to use to enclose a rectangular space for a garden. This table shows the width, length, and area of some of the fences you could build.

Width (m)	0	1	3.5	5	6	8	10.5	12
Length (m)	12	11	8.5	7	6	4	1.5	0
Area (sq. m)	0	11	29.75	35	36	32	15.75	0

Notice that width values of 0 and 12 give areas of 0. Enter the width values into list L1 and enter the area values into list L2.

Make a graph of the points (x, y), where x is the width of the rectangle and y is the area. Plot additional points to help you determine the shape of the graph. The points appear to fall in the shape of a parabola. Any real-number value between 0 and 12 makes sense for the width, so you can connect the points with a smooth curve.

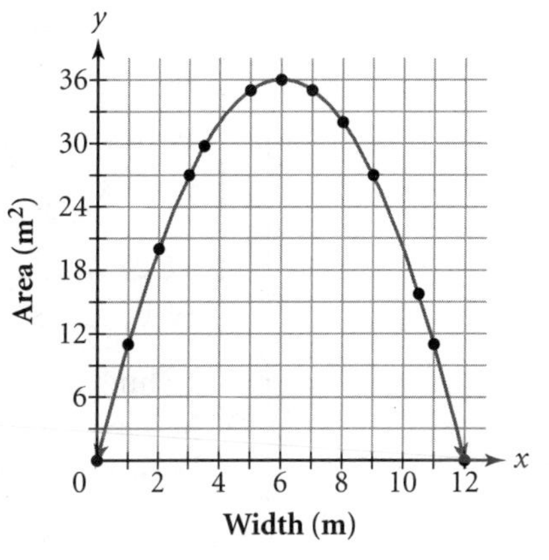

The graph appears to reach its highest point at (6, 36), indicating that a rectangle with width 6 meters has the greatest area, 36 square meters. The length of this rectangle is also 6 meters, so the rectangle is a square.

A garden with width 2 meters has length 10 meters. A garden with width 4.3 meters has length 7.7 meters. In general, if x is the width of the garden, the length is $12 - x$ and the equation for the area y is $y = x(12 - x)$. In the same window, graph this equation and plot the (L1, L2) values.

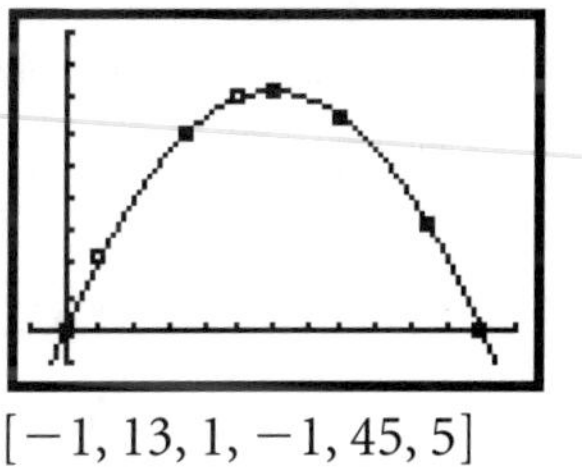

[−1, 13, 1, −1, 45, 5]

By tracing the graph of $y = x(12 - x)$ to find the coordinates of the vertex, you can verify that the square with side length 6 has the maximum area.

(continued)

The points where a graph crosses the x-axis are called the **x-intercepts.** The x-intercepts of the graph above are (0, 0) and (12, 0). This indicates that the rectangle has no area if the width is 0 meter or 12 meters. (Note: When naming an x-intercept, sometimes just the x-coordinate, rather than the ordered pair, is given. So, we might say that the x-intercepts of the graph above are 0 and 12.)

If you repeat this process for different perimeters (total fence lengths), you will find that the rectangle with the greatest area is always a square.

The x-coordinates of the x-intercepts of a graph are the solutions of the equation $f(x) = 0$. These solutions are called the **roots** of the function $y = f(x)$. For the equation in the investigation, the roots are 0 and 12, the width values that make the area equal zero.

Read the rest of the lesson in your book. Example A shows you how to use your calculator to estimate the roots of a quadratic function. Example B shows how to find the vertex and the **line of symmetry** of a parabola based on its equation and then how to use this information to rewrite the equation in **vertex form.** Here is another example.

EXAMPLE Find the line of symmetry and the vertex of the parabola $y = x^2 + 9x + 14$. Then, write the equation in vertex form $y = a(x - h)^2 + k$.

▶ ***Solution*** You can use a calculator graph to find that the roots are -7 and -2.

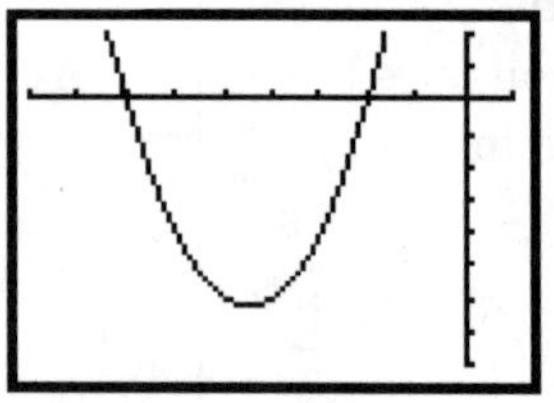

$[-9, 1, 1, -8, 2, 1]$

The line of symmetry is $x = -4.5$, the vertical line halfway between the x-intercepts. The vertex lies on the line of symmetry, so it has an x-coordinate of -4.5. To find its y-coordinate, substitute -4.5 for x in the equation.

$$y = (-4.5)^2 + 9(-4.5) + 14$$

$$= 20.25 - 40.5 + 14$$

$$= -6.25$$

So, the vertex is $(-4.5, -6.25)$.

The graph is a transformation of the parent function, $f(x) = x^2$. Because the vertex is $(-4.5, -6.25)$, there is a translation left 4.5 units and down 6.25 units, so the equation is of the form $y = a(x + 4.5)^2 - 6.25$. If you graph $y = (x + 4.5) - 6.25$ in the same window as the original equation, you will find that the graphs are the same. So, $y = (x + 4.5)^2 - 6.25$ is the vertex form of $y = x^2 + 9x + 14$.

CONDENSED LESSON

10.3 From Vertex to General Form

In this lesson you will

- draw diagrams to **square expressions**
- draw diagrams to write **trinomials** as squares of expressions
- convert a quadratic equation from **vertex form** to **general form**

In the general form of a quadratic equation, $y = ax^2 + bx + c$, the right side is the sum of three terms. A **term** is an algebraic expression that represents only multiplication and division between variables and constants. A sum of terms with positive integer exponents is called a **polynomial.** Read about polynomials on page 544 of your book.

Example A gives you practice identifying polynomials. Read and follow along with this example. Then, read the text about *like terms* that follows the example.

Investigation: Sneaky Squares

The diagram in Step 1 in your book expresses 7^2 as $(3 + 4)^2$. The sum of the areas of the inner rectangles is $9 + 2(12) + 16 = 49$. This verifies that $7^2 = (3 + 4)^2$.

Now, draw and label a similar diagram for each expression in Step 2. Here are the results for parts a and b.

a. $(5 + 3)^2 = 25 + 2(15) + 9 = 64$

	5	3
5	25	15
3	15	9

b. $(4 + 2)^2 = 16 + 2(8) + 4 = 36$

	4	2
4	16	8
2	8	4

You can use similar diagrams to show the squares of differences, even though negative lengths are involved. The diagram before Step 3 in your book shows 7^2 as $(10 - 3)^2$. Draw and label a diagram for each expression in Step 3. Here are the results for parts a and b.

a. $(5 - 2)^2 = 25 + 2(-10) + 4 = 9$

	5	−2
5	25	−10
−2	−10	4

b. $(7 - 3)^2 = 49 + 2(-21) + 9 = 16$

	7	−3
7	49	−21
−3	−21	9

The expressions in Step 4 involve variables. Draw and label a diagram for each expression. Combine like terms to express each answer as a trinomial. The results for parts a and b are on the next page.

(continued)

a. $(x + 5)^2 = x^2 + 10x + 25$

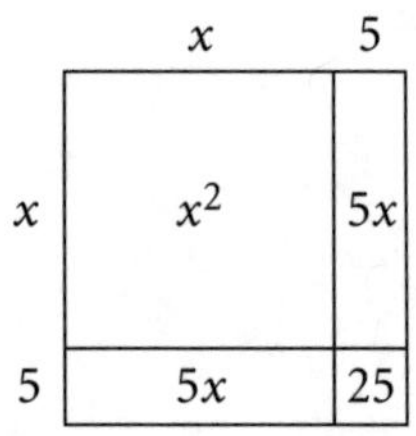

b. $(x - 3)^2 = x^2 - 6x + 9$

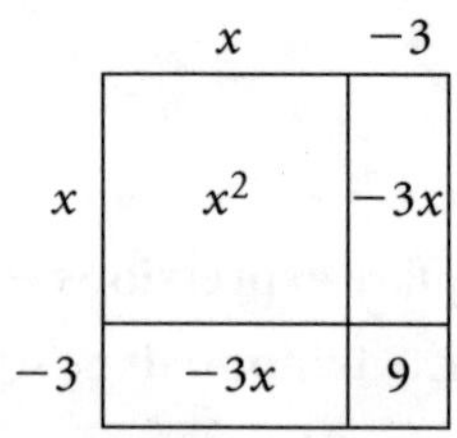

For certain types of trinomials, you can make a rectangular diagram and then write an equivalent expression in the form $(x + h)^2$. Try this for the expressions in Step 5. Here are the results for parts a and b. In each case, the diagram is divided into a square with area equal to the first term, a square with area equal to the last term, and two rectangles, each with area equal to *half* the middle term.

a. $x^2 + 6x + 9 = (x + 3)^2$

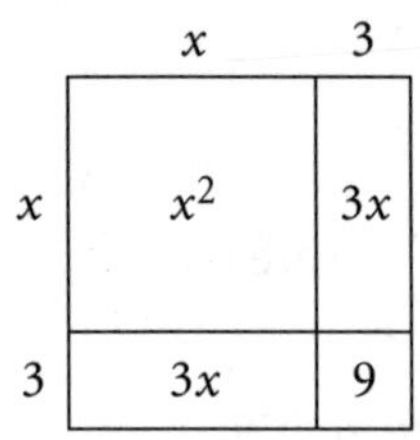

b. $x^2 - 10x + 25 = (x - 5)^2$

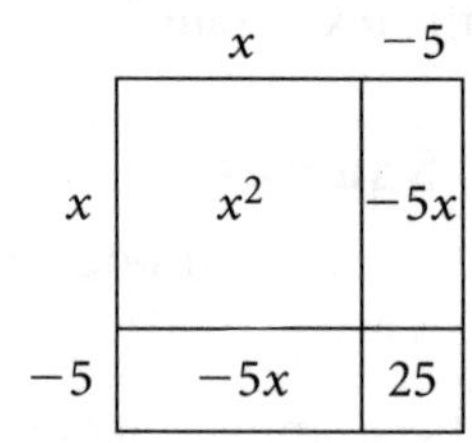

You can use the results from Step 5 to solve the equations in Step 6. Here are the results for parts a and b.

$$\begin{aligned} x^2 + 6x + 9 &= 49 \\ (x + 3)^2 &= 49 \\ \sqrt{(x + 3)^2} &= \sqrt{49} \\ x + 3 &= \pm 7 \\ x &= -3 \pm 7 \\ x &= 4 \text{ or } x = -3 \end{aligned}$$

$$\begin{aligned} x^2 - 10x + 25 &= 81 \\ (x - 5)^2 &= 81 \\ \sqrt{(x - 5)^2} &= \sqrt{81} \\ x - 5 &= \pm 9 \\ x &= 5 \pm 9 \\ x &= 14 \text{ or } x = -4 \end{aligned}$$

Numbers such as 25 are called **perfect squares** because they are the squares of integers, in this case, 5 and -5. The trinomial $x^2 - 10x + 25$ is also called a perfect square because it is the square of $x - 5$. If the coefficient of the x^2-term is 1, then a trinomial is a perfect square if the last term is the square of half the coefficient of the x-term. Use this idea to identify the perfect squares in Step 7. Here are the results.

a. Because 49 is equal to the square of half of 14, this is a perfect-square trinomial: $x^2 + 14x + 49 = (x + 7)^2$.

b. Because 81 is equal to the square of half of -18, this is a perfect-square trinomial: $x^2 - 18x + 81 = (x - 9)^2$.

c. This is not a perfect-square trinomial because 25 is not equal to the square of half of 20.

d. This is not a perfect-square trinomial because -36 is not equal to the square of half of -12.

You can use your skills at squaring binomials to convert equations from vertex form to general form. This is illustrated in Example B in your text. Read that example and the text that follows it.

CONDENSED LESSON

10.4 Factored Form

In this lesson you will

- work with quadratic equations in **factored form**
- learn the connection between the factored form of a quadratic equation and the equation's roots
- write the equation of a parabola in three different forms

You have worked with quadratic equations given in vertex form and in general form. In this lesson you will learn about the **factored form** of a quadratic equation.

Investigation: Getting to the Root of the Matter

Steps 1–4 In the same window, graph $y = x + 3$ and $y = x - 4$.

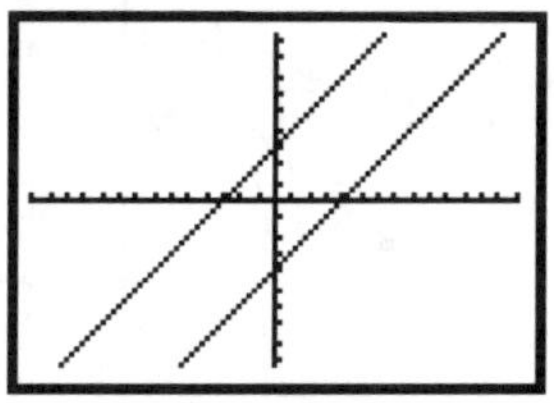

$[-14.1, 14.1, 1, -9.3, 9.3, 1]$

The x-intercept of $y = x + 3$ is -3. The x-intercept of $y = x - 4$ is 4.

Now, in the same window, graph $y = (x + 3)(x - 4)$. The graph is a parabola.

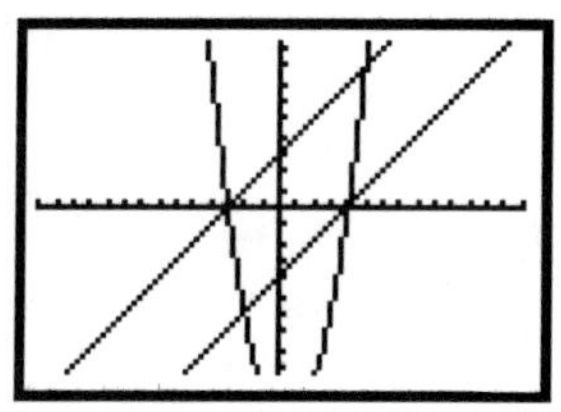

The x-intercepts of the parabola are -3 and 4—the x-intercepts of $y = x + 3$ and $y = x - 4$, respectively. This makes sense because the product $(x + 3)(x - 4)$ is zero when $x + 3$ is zero or when $x - 4$ is zero.

You can use a rectangular diagram to expand the expression $(x + 3)(x - 4)$ and then rewrite $y = (x + 3)(x - 4)$ as $y = x^2 - x - 12$. Verify that the two equations are equivalent by graphing both on the same axes. Because the equations are identical, you know that the *roots* of $y = x^2 - x - 12$ are -3 and 4.

	x	-4
x	x^2	$-4x$
3	$3x$	-12

Steps 5–8 If you are given a quadratic equation in general form, you can sometimes use a rectangular diagram to rewrite it in factored form and then find its roots.

Consider the equation $y = x^2 + 5x + 6$. The diagram shows that you can rewrite the left side as $(x + 3)(x + 2)$. So, in factored form, the equation is $y = (x + 3)(x + 2)$. Use a calculator graph or table to verify that $y = x^2 + 5x + 6$ and $y = (x + 3)(x - 2)$ are equivalent.

	x	2
x	x^2	$2x$
3	$3x$	6

From the factored form, you can see that the roots of $y = x^2 + 5x + 6$ are -3 and -2.

(continued)

Now, use rectangle diagrams to rewrite each equation in Step 8 in factored form and find its roots. Here are the results.

a. $y = (x - 5)(x - 2)$; roots: 5 and 2

b. $y = (x + 8)(x - 2)$; roots: -8 and 2

c. $y = (x - 6)(x + 8)$; roots: 6 and -8

d. $y = (x - 7)(x - 4)$; roots 7 and 4

Now, read the text after the investigation on page 552 in your book, which summarizes the three forms of a quadratic equation. Then, read the example, which shows you how to write an equation of a parabola in all three forms, and read the text that follows the example. Here is an additional example.

EXAMPLE Write the equation for this parabola in vertex form, general form, and factored form.

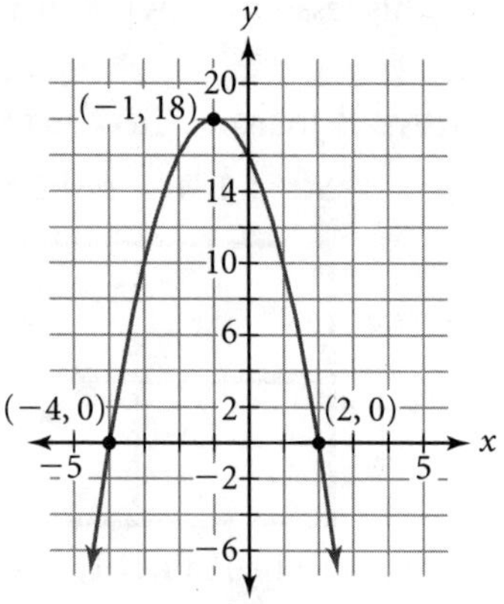

Solution From the graph, you can see that the x-intercepts are -4 and 2. So, the factored form contains the binomial factors $(x + 4)$ and $(x - 2)$. You can also see that, because the parabola is "upside down," the coefficient of x^2 must be negative.

If you graph $y = -(x + 4)(x - 2)$ on your calculator, you'll see that it has the same x-intercepts as the graph above, but a different vertex.

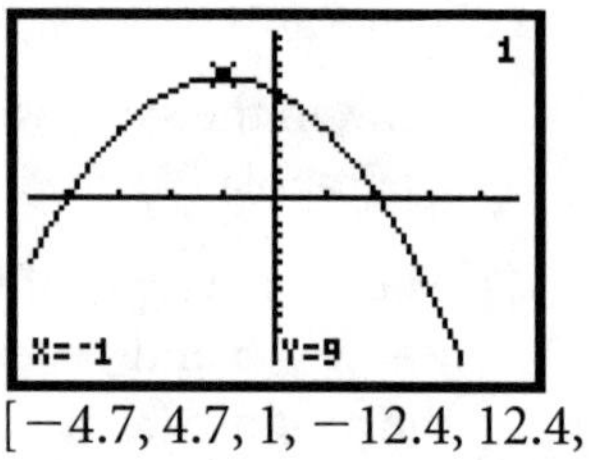

$[-4.7, 4.7, 1, -12.4, 12.4, 1]$

The vertex of $y = -(x + 4)(x - 2)$ is $(-1, 9)$, while the vertex of the original parabola is $(-1, 18)$. So, the original parabola is a vertical stretch of the graph of $y = -(x + 4)(x - 2)$ by a factor of $\frac{18}{9}$, or 2. This means the factored form is $y = -2(x + 4)(x - 2)$. Graph this equation on your calculator to verify that its graph looks like the original graph.

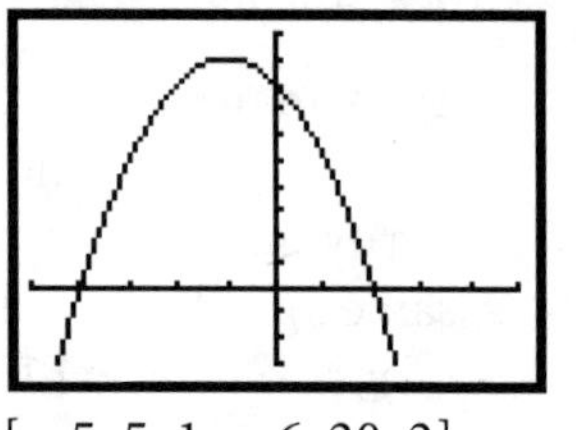

$[-5, 5, 1, -6, 20, 2]$

You now know that the value of a is -2 and the vertex is $(-1, 18)$, so you can write the vertex form of the equation, $y = -2(x + 1)^2 + 18$. To get the general form, expand either the factored or vertex form.

$y = -2(x + 1)^2 + 18$	The vertex form.
$y = -2(x^2 + 2x + 1) + 18$	Use a rectangle diagram to expand $(x + 1)^2$.
$y = -2x^2 - 4x - 2 + 18$	Use the distributive property.
$y = -2x^2 - 4x + 16$	Combine like terms.

CONDENSED LESSON

10.6 Completing the Square

In this lesson you will

- solve quadratic equations by **completing the square**
- rewrite quadratic equations in vertex form by completing the square
- work with a quadratic equation that has **no real-number solutions**

You can find approximate solutions to quadratic equations by using calculator graphs and tables. If you are able to write the equation in factored form or in vertex form, you can use symbolic methods to find the exact solutions. In this lesson you'll learn a symbolic method called **completing the square** that you can use to find exact solutions to any quadratic equation in the general form, $y = ax^2 + bx + c$.

Investigation: Searching for Solutions

Complete each rectangular diagram in Step 1 so that it is a square. Here are the results for parts a, b, and d.

a.

	x	2
x	x^2	$2x$
2	$2x$	4

b.

	x	-3
x	x^2	$-3x$
-3	$-3x$	9

d.

	x	$\frac{b}{2}$
x	x^2	$\frac{b}{2}x$
$\frac{b}{2}$	$\frac{b}{2}x$	$\frac{b^2}{4}$

Using the diagrams, you can write equations of the form $x^2 + bx + c = (x + h)^2$. Here are the equations for parts a, b, and d. Write a similar equation for part c.

a. $x^2 + 4x + 4 = (x + 2)^2$ **b.** $x^2 - 6x + 9 = (x - 3)^2$ **d.** $x^2 + bx + \frac{b^2}{4} = \left(x + \frac{b}{2}\right)^2$

Notice that the operations on the right side of each equation can be "undone" to get x. For example, you can undo $(x + 2)^2$ by taking the square root of $x + 2$ and then subtracting 2. You cannot undo the operations on the left side of the equation to end up with x, so the $(x + h)^2$ form is used to solve problems.

If you assume the area of each square in Step 1 is 100, you can write equations that can be solved by undoing the order of operations. Here are the equations for parts a and b. Write a similar equation for part c.

a. $(x + 2)^2 = 100$ **b.** $(x - 3)^2 = 100$

Here are the symbolic solutions of the equations above. Write a similar solution of the equation for part c.

a.

$$(x + 2)^2 = 100$$
$$\sqrt{(x + 2)^2} = \sqrt{100}$$
$$x + 2 = \pm 10$$
$$x = -2 \pm 10$$
$$x = -12 \text{ or } x = 8$$

b.

$$(x - 3)^2 = 100$$
$$\sqrt{(x - 3)^2} = \sqrt{100}$$
$$x - 3 = \pm 10$$
$$x = 3 \pm 10$$
$$x = 13 \text{ or } x = -7$$

(continued)

When an equation cannot easily be rewritten in factored or vertex form, you can solve it by completing the square. The example below illustrates this method. Make sure you understand what is happening at each step.

$x^2 + 6x - 1 = 0$	The original equation.
$x^2 + 6x = 1$	Add 1 to both sides.
$x^2 + 6x + 9 = 1 + 9$	Add 9 to both sides, making the left side a perfect-square trinomial.
$(x + 3)^2 = 10$	Rewrite the trinomial as a squared binomial.
$x + 3 = \pm\sqrt{10}$	Take the square root of both sides.
$x = -3 \pm \sqrt{10}$	Add -3 to both sides.

The solutions are $-3 + \sqrt{10}$, which is about 0.162, and $-3 - \sqrt{10}$, which is about -6.162. You can check these solutions by graphing $y = x^2 + 6x - 1$ and finding the x-intercepts or by making a table and finding the x-values that correspond to the y-value 0.

Here are the steps for solving $x^2 + 8x - 5 = 0$.

$x^2 + 8x - 5 = 0$	The original equation.
$x^2 + 8x = 5$	Add 5 to both sides.
$x^2 + 8x + 16 = 5 + 16$	Add 16 to both sides, making the left side a perfect-square trinomial.
$(x + 4)^2 = 21$	Rewrite the trinomial as a squared binomial.
$x + 4 = \pm\sqrt{21}$	Take the square root of both sides.
$x = -4 \pm \sqrt{21}$	Add -4 to both sides.

The solutions are $-4 + \sqrt{21}$, which is about 0.583, and $-4 - \sqrt{21}$, which is about -8.583.

The key to solving a quadratic equation by completing the square is to express one side as a perfect-square trinomial. In the investigation the equations were in the form $y = 1x^2 + bx + c$. Example A in your book shows how to solve a quadratic equation when the coefficient of x^2 is not 1. Read this example carefully. Then, read Example B, which shows how to complete the square to convert an equation to vertex form.

In the solution of $2(x + 2)^2 + 3 = 0$ in Example B, the final step is $x = -2 \pm \sqrt{-\frac{3}{2}}$. Because a negative number does not have a real-number square root, the equation has no real-number solutions. The graph of the equation $y = 2(x + 2)^2 + 3$ on page 563 can help you understand why there are no solutions. The graph does not cross the x-axis, so there is no real-number value of x that makes $2(x + 2)^2 + 3 = 0$ true.

CONDENSED LESSON

10.7 The Quadratic Formula

In this lesson you will

- see how the **quadratic formula** is derived
- use the quadratic formula to solve equations

If a quadratic equation is given in vertex form or if it has no x-term, you can solve it by undoing the operations or by keeping the equation balanced. If the equation is in factored form, you can solve it by finding the x-values that make the factors equal to zero. If the equation is a perfect-square trinomial, you can factor it and then find the solutions. Look at the six equations given at the beginning of the lesson, and think about how you would solve each one.

In the previous lesson you learned how to solve quadratic equations by completing the square. Unfortunately, this method can sometimes be messy. For example, try solving $-4.9x^2 + 5x - \frac{16}{3} = 0$ by completing the square.

Consider the general quadratic equation $ax^2 + bx + c = 0$. In this lesson you will see how completing the square in this general case leads to a formula that can be used to solve any quadratic equation.

Investigation: Deriving the Quadratic Formula

Consider the equation $2x^2 + 3x - 1 = 0$. This equation is in the general form, $ax^2 + bx + c = 0$, where $a = 2$, $b = 3$, and $c = -1$.

In this investigation, we will show the steps for solving the general equation, $ax^2 + bx + c = 0$, on the left and the steps for solving the particular equation, $2x^2 + 3x - 1 = 0$, on the right.

First, group all the variable terms on the left side of the equation.

$ax^2 + bx = -c$	$2x^2 + 3x = 1$

To complete the square, the coefficient of x^2 must be 1. So, divide both sides of the equation by the value of a.

$x^2 + \frac{b}{a}x = -\frac{c}{a}$	$x^2 + \frac{3}{2}x = \frac{1}{2}$

Complete the square of the left side of the equation by adding the square of half the coefficient of x. Add this same value to the right side.

$x^2 + \frac{b}{a}x + \left(\frac{b}{2a}\right)^2 = \left(\frac{b}{2a}\right)^2 - \frac{c}{a}$	$x^2 + \frac{3}{2}x + \left(\frac{3}{4}\right)^2 = \left(\frac{3}{4}\right)^2 + \frac{1}{2}$

Rewrite the trinomial on the left side of the equation as a squared binomial. On the right side, rewrite the fractions with a common denominator.

$\left(x + \frac{b}{2a}\right)^2 = \frac{b^2}{4a^2} - \frac{4ac}{4a^2}$	$\left(x + \frac{3}{4}\right)^2 = \frac{9}{16} + \frac{8}{16}$

Take the square root of both sides.

$x + \frac{b}{2a} = \pm\frac{\sqrt{b^2 - 4ac}}{\sqrt{4a^2}}$	$x + \frac{3}{4} = \pm\frac{\sqrt{9 + 8}}{\sqrt{16}}$

(continued)

Get x by itself on the left side.

$$x = -\frac{b}{2a} \pm \frac{\sqrt{b^2 - 4ac}}{2a} \qquad\qquad x = -\frac{3}{4} \pm \frac{\sqrt{17}}{4}$$

Express the solutions as fractions.

$$x = \frac{-b + \sqrt{b^2 - 4ac}}{2a} \qquad\qquad x = \frac{-3 + \sqrt{17}}{4}$$

$$\text{or } x = \frac{-b - \sqrt{b^2 - 4ac}}{2a} \qquad\qquad \text{or } x = \frac{-3 - \sqrt{17}}{4}$$

In decimal form, the solutions to $2x^2 + 3x - 1 = 0$ are approximately 0.281 and -1.781.

Look at the solutions to the general form. Notice that, for the solutions to be real numbers, the value of a cannot be zero (because division by zero is undefined) and the value of $b^2 - 4ac$ must be greater than or equal to zero (because negative numbers do not have real square roots).

The quadratic formula, $x = \frac{-b \pm \sqrt{b^2 - 4ac}}{2a}$, gives the solutions to a quadratic equation written in the general form, $ax^2 + bx + c = 0$. To use the formula, you need only know the values of a, b, and c. The example in your book illustrates how to use the formula. Here is another example.

EXAMPLE | Use the quadratic formula to solve $2x^2 - 9 = x$.

▶ Solution

First, write the equation in standard form by subtracting x from both sides. The result is $2x^2 - x - 9 = 0$. For this equation, $a = 2$, $b = -1$, and $c = -9$. Now, substitute these values into the quadratic formula.

$$x = \frac{-b \pm \sqrt{b^2 - 4ac}}{2a}$$

$$x = \frac{-(-1) \pm \sqrt{(-1)^2 - 4(2)(-9)}}{2(2)}$$

$$x = \frac{1 \pm \sqrt{1 - (-72)}}{4}$$

$$x = \frac{1 \pm \sqrt{73}}{4}$$

So, the solutions are $\frac{1 + \sqrt{73}}{4}$, or about 2.386, and $\frac{1 - \sqrt{73}}{4}$, or about -1.886.

Read the rest of the lesson in your book.

CONDENSED LESSON

10.8 Cubic Functions

In this lesson you will

- determine whether given numbers are **perfect cubes**
- discover the connection between the factored form of a **cubic equation** and its graph
- write equations for **cubic functions** based on their graphs

Read the text at the bottom of page 572 of your book. It explains that the cubing function, $f(x) = x^3$, models the volume of a cube with edge length x. Functions in the family with parent function $f(x) = x^3$ are called *cubic functions.*

The volume of a cube with edge length 5 is 5^3, or 125. The number 125 is a **perfect cube** because it is equal to an integer cubed (that is, raised to the third power). The number 5 is called the **cube root** of 125. You can express this by writing $5 = \sqrt[3]{125}$.

EXAMPLE A Determine which numbers are perfect cubes.

a. 24,389 **b.** 1,428 **c.** 270 **d.** 4,913

▸ Solution Find the cube root of each number. (See **Calculator Note 10B.**)

a. $\sqrt[3]{24{,}389} = 29$ **b.** $\sqrt[3]{1{,}428} \approx 11.261$

c. $\sqrt[3]{270} \approx 6.463$ **d.** $\sqrt[3]{4{,}913} = 17$

So, 24,389 and 4,913 are perfect cubes because their cube roots are integers.

The graph of the parent function $y = x^3$ is shown on page 572 of your book. You can use what you know about transformations to write equations for other cubic functions.

EXAMPLE B Write an equation for each graph.

a.

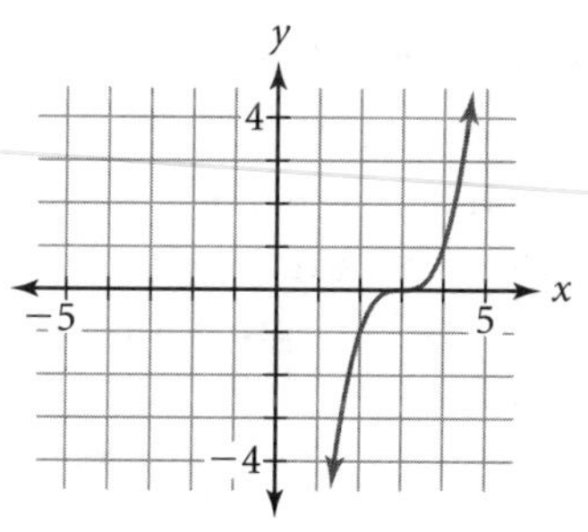

b.

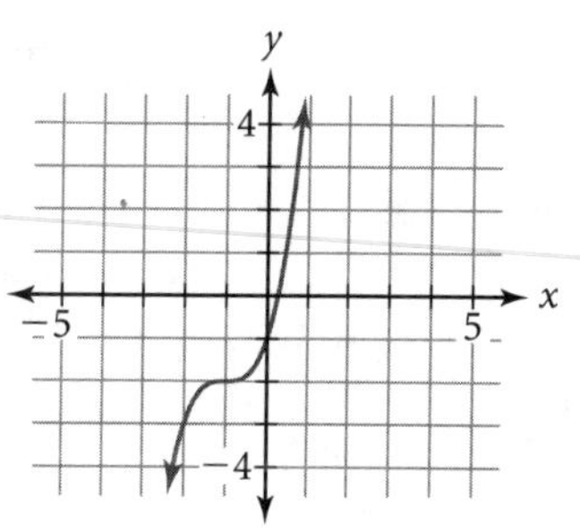

▸ Solution

a. The graph is a translation of the graph of $y = x^3$ right 3 units. So, the equation is $y = (x - 3)^3$.

b. The graph is a translation of the graph of $y = x^3$ left 1 unit and down 2 units. So, the equation is $y = (x + 1)^3 - 2$.

(continued)

Investigation: Rooting for Factors

In this investigation you'll discover the connection between the factored form of a cubic equation and its graph.

Steps 1–3 List the x-intercepts for each graph in Step 1. Here are the results.

Graph A: $-2, 1, 2$ Graph B: $-2, 1, 2$ Graph C: $-1, 0, 2$

Graph D: $-1, 0, 2$ Graph E: $-2, 1, 3$ Graph F: $-3, -1, 2$

Use tables and graphs to help you match the equations in Step 2 with the graphs in Step 1. Here are the results.

a. Graph F **b.** Graph C **c.** Graph A

d. Graph D **e.** Graph B **f.** Graph E

The x-intercepts are the roots of the equation. So, if the graph of a cubic equation has x-intercept a, the factored form of the equation includes the factor $(x - a)$. For example, Graph A has x-intercepts at -2, 1, and 2, and its equation includes the factors $(x + 2)$, $(x - 1)$, and $(x - 2)$.

Step 4 Now, you will find an equation for the graph in Step 4. It has x-intercepts -3, -2, and 2. The equation $y = (x + 3)(x + 2)(x - 2)$ has the same x-intercepts. Here is its graph.

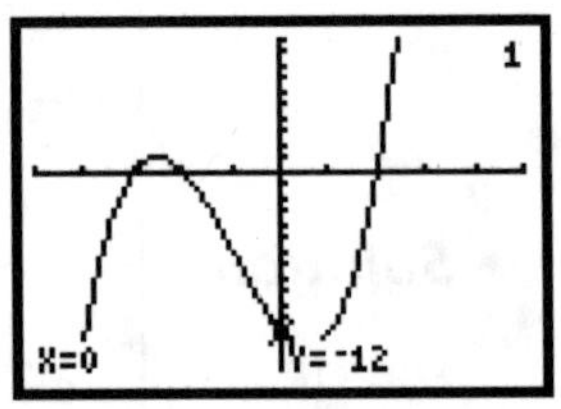

$[-5, 5, 1, -15, 10, 1]$

Now, make adjustments to the equation until the graph looks like the one in Step 4. The point $(0, -12)$ corresponds to the point $(0, 6)$ in the original graph. So, you need to reflect the graph over the x-axis and apply a vertical shrink by a factor of 0.5. The equation then becomes $y = -0.5(x + 3)(x + 2)(x - 2)$. If you graph this equation on your calculator, you'll see that the result matches the graph in Step 4.

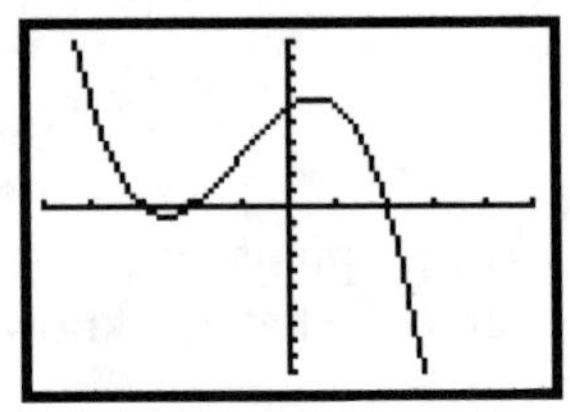

$[-5, 5, 1, -10, 10, 1]$

Read Example C in your book, which shows how to find an equation for another cubic graph.

If you are given a cubic equation in general form, you can graph it and then use the x-intercepts to help you write the equation in factored form. The text after Example C shows how to rewrite $y = x^3 - 3x + 2$ in factored form. The graph of this equation touches the x-axis at $x = 1$ but doesn't actually pass through the axis at this point. This indicates a *double root.* This means that the factor $x - 1$ appears twice in the equation. The factored form of the equation is $y = (x + 2)(x - 1)^2$.

CONDENSED LESSON

11.1 Parallel and Perpendicular

In this lesson you will

- learn the meaning of **parallel** and **perpendicular**
- discover how the **slopes of parallel and perpendicular lines** are related
- use slopes to help **classify figures** in the coordinate plane

Parallel lines are lines in the same plane that never intersect.

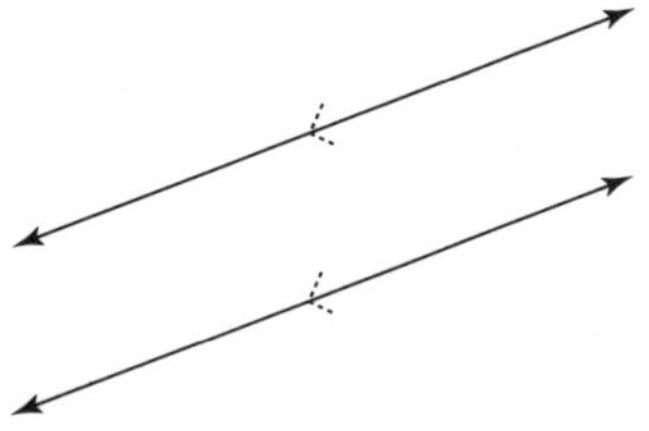

Perpendicular lines are lines in the same plane that intersect at a right angle.

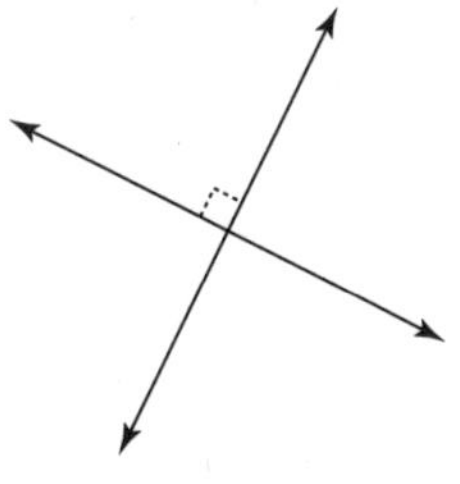

We show a small box in one of the angles to show that the lines are perpendicular.

Investigation: Slopes

The opposite sides of a rectangle are parallel, and the adjacent sides are perpendicular. By examining rectangles drawn on a coordinate grid, you can discover how the slopes of parallel and perpendicular lines are related.

Step 1 gives the vertices of four rectangles. Here is the rectangle with the vertices given in part a.

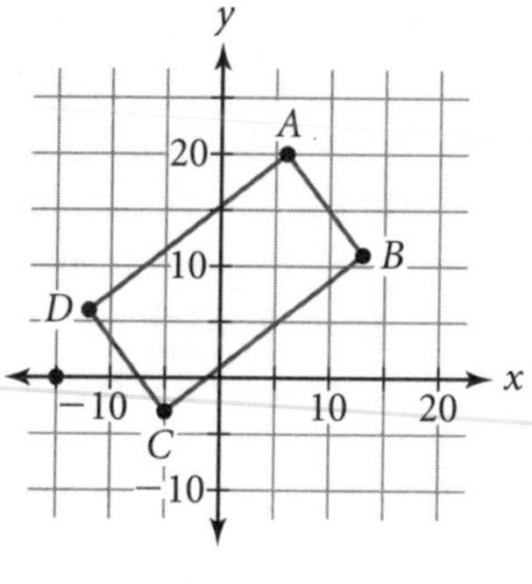

Find the slope of each side of the rectangle. You should get these results. (Note: The notation $\overline{AB}$ means "segment AB.")

Slope of $\overline{AD}$: $\frac{7}{9}$ Slope of $\overline{AB}$: $-\frac{9}{7}$

Slope of $\overline{BC}$: $\frac{7}{9}$ Slope of $\overline{DC}$: $-\frac{9}{7}$

Notice that the slopes of the parallel sides $\overline{AD}$ and $\overline{BC}$ are the same and that the slopes of the parallel sides $\overline{AB}$ and $\overline{DC}$ are the same.

Recall that, to find the **reciprocal** of a fraction, you exchange the numerator and the denominator. For example, the reciprocal of $\frac{3}{4}$ is $\frac{4}{3}$. The product of a number and its reciprocal is 1. Look at the slopes of the perpendicular sides $\overline{AD}$ and $\overline{DC}$. The slope of $\overline{DC}$ is the *negative reciprocal* of the slope of $\overline{AD}$. The product of the slopes, $\frac{7}{9}$ and $-\frac{9}{7}$, is -1. You'll find this same relationship for any pair of perpendicular sides of the rectangle.

(continued)

Now, choose another set of vertices from Step 1, and find the slopes of the sides of the rectangle. You should find the same relationships among the slopes of the sides. In fact, any two parallel lines have the same slope, and any two perpendicular lines have slopes that are negative reciprocals.

A **right triangle** has one right angle. The sides that form the right angle are called **legs,** and the side opposite the right angle is called the **hypotenuse.** If a triangle is drawn on a coordinate grid, you can use what you know about slopes of perpendicular lines to determine if it is a right triangle. This is demonstrated in Example A in your book. Here is another example.

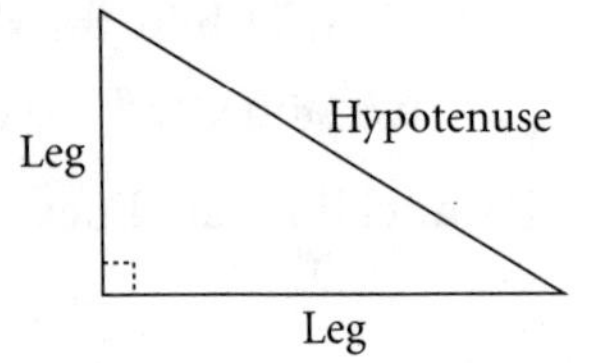

EXAMPLE Decide whether this triangle is a right triangle.

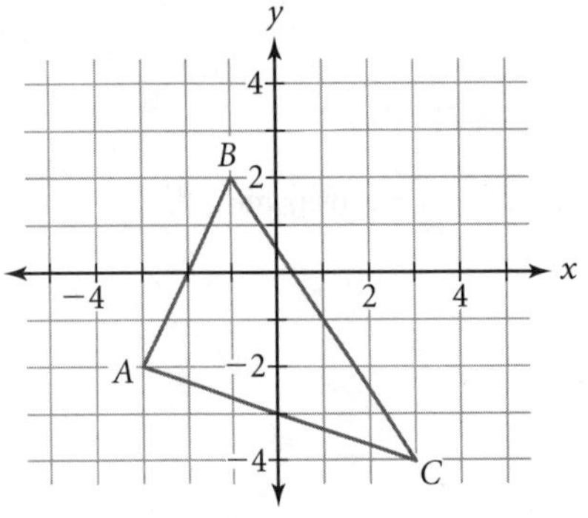

▶ ***Solution*** This triangle has vertices $A(-3, -2)$, $B(-1, 2)$, and $C(3, -4)$. Angles B and C are clearly not right angles, but angle A might be. To check, find the slopes of $\overline{AB}$ and $\overline{AC}$:

Slope $\overline{AB}$: $\dfrac{2-(-2)}{-1-(-3)} = \dfrac{4}{2} = 2$ Slope $\overline{AC}$: $\dfrac{-4-(-2)}{3-(-3)} = \dfrac{-2}{6} = -\dfrac{1}{3}$

The slopes, 2 and $-\frac{1}{3}$, are not negative reciprocals, so the sides are not perpendicular. Because none of the angles are right angles, the triangle is not a right triangle.

A **trapezoid** is a quadrilateral with one pair of opposite sides that are parallel and one pair of opposite sides that are not parallel. A trapezoid with a right angle is a **right trapezoid.** Every right trapezoid must have two right angles because opposite sides are parallel. Here are some examples of trapezoids.

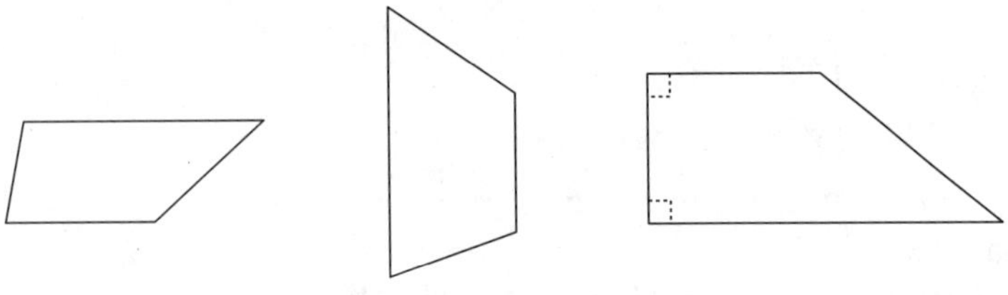

To determine whether a quadrilateral drawn on a coordinate grid is a trapezoid that is not also a parallelogram, you need to check that two of the opposites sides have the same slope and the other two opposite sides have different slopes. To decide whether the trapezoid is a right trapezoid, you also need to check that the slopes of two adjacent sides are negative reciprocals. This is illustrated in Example B in your book.

CONDENSED LESSON

11.2 Finding the Midpoint

In this lesson you will

- discover the **midpoint formula**
- use the midpoint formula to find midpoints of segments
- write equations for **medians** of triangles and **perpendicular bisectors** of segments

The **midpoint** of a line segment is the middle point—that is, the point halfway between the endpoints. The text on page 588 of your book explains that finding midpoints is necessary for drawing the **median** of a triangle and the **perpendicular bisector** of a line segment. Read this text carefully.

Investigation: In the Middle

This triangle has vertices $A(1, 2)$, $B(5, 2)$, and $C(5, 7)$.

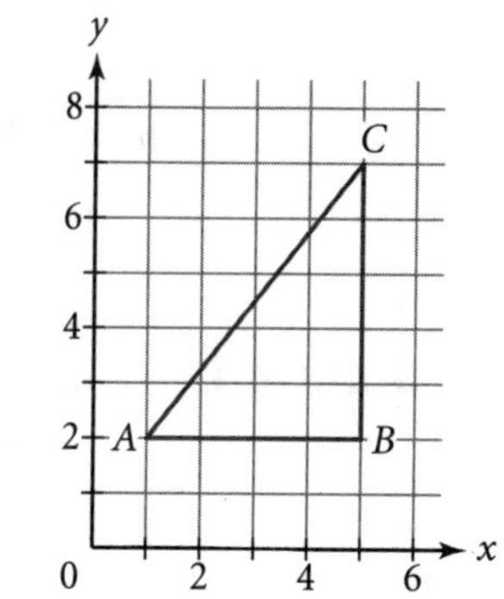

The midpoint of $\overline{AB}$ is $(3, 2)$. Notice that the x-coordinate of this point is the average of the x-coordinates of the endpoints.

The midpoint of $\overline{BC}$ is $(5, 4.5)$. Notice that the y-coordinate of this point is the average of the y-coordinates of the endpoints.

The midpoint of $\overline{AC}$ is $(3, 4.5)$. Notice that the x-coordinate of this point is the average of the x-coordinates of the endpoints and that the y-coordinate is the average of the y-coordinates of the endpoints.

Segment DE has endpoints $D(2, 5)$ and $E(7, 11)$. The midpoint of $\overline{DE}$ is $(4.5, 8)$. The x-coordinate of this point is the average of the x-coordinates of the endpoints, and the y-coordinate is the average of the y-coordinates of the endpoints.

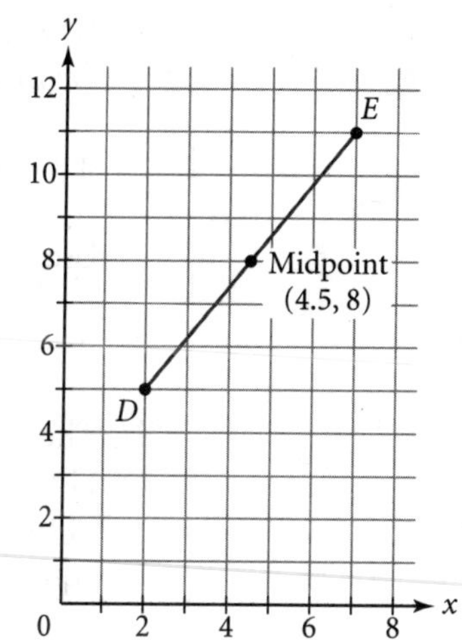

Use the idea of averaging the coordinates of the endpoints to find the midpoint of the segment between each pair of points. For Step 8, you should get these results.

a. midpoint of $\overline{FG}$: $(-2.5, 28)$

b. midpoint of $\overline{HJ}$: $(-1, -2)$

The technique used in the investigation to find the midpoint of a segment is known as the **midpoint formula.** If the endpoints of a segment have coordinates (x_1, y_1) and (x_2, y_2), the midpoint of the segment has coordinates

$$\left(\frac{x_1 + x_2}{2}, \frac{y_1 + y_2}{2}\right)$$

(continued)

The example in your book shows how to find equations for a median of a triangle and the perpendicular bisector of one of its sides. Below is another example.

EXAMPLE This triangle has vertices $A(-2, 2)$, $B(2, 4)$, and $C(1, -3)$.

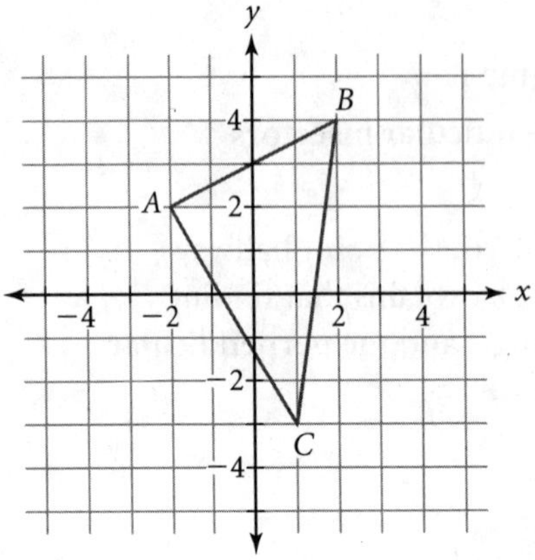

a. Write the equation of the median from vertex A.

b. Write the equation of the perpendicular bisector of $\overline{BC}$.

▶ ***Solution***

a. The median from vertex A goes to the midpoint of $\overline{BC}$. So, find the midpoint of $\overline{BC}$.

midpoint of $\overline{BC}$: $\left(\frac{2+1}{2}, \frac{4+(-3)}{2}\right) = (1.5, 0.5)$

Now, use the coordinates of vertex A and the midpoint to find the slope of the median.

slope of median: $\frac{0.5-2}{1.5-(-2)} = \frac{-1.5}{3.5} = -\frac{3}{7}$

Use the coordinates of the midpoint and the slope to find the equation.

$y = 0.5 - \frac{3}{7}(x - 1.5)$

b. The perpendicular bisector of $\overline{BC}$ goes through the midpoint of $\overline{BC}$, which is $(1.5, 0.5)$ and is perpendicular to $\overline{BC}$. The slope of $\overline{BC}$ is $\frac{-3-4}{1-2}$, or 7, so the slope of the perpendicular bisector is the negative reciprocal of 7, or $-\frac{1}{7}$. Write the equation, using this slope and the coordinates of the midpoint.

$y = 0.5 - \frac{1}{7}(x - 1.5)$

CONDENSED LESSON

11.3 Squares, Right Triangles, and Areas

In this lesson you will

- find the **areas of polygons** drawn on a grid
- find the **area** and **side length of squares** drawn on a grid
- draw a segment of a given length by drawing a square with the square of the length as area

Example A in your book shows you how to find the length of a rectangle and a right triangle. Example B demonstrates how to find the area of a tilted square by drawing a square with horizontal and vertical sides around it. Read both examples carefully.

Investigation: What's My Area?

Step 1 Find the area of each figure in Step 1. You should get these results.

a. 1 square unit **b.** 5 square units **c.** 6 square units

d. 2 square units **e.** 8 square units **f.** 3 square units

g. 6 square units **h.** 6 square units **i.** 10.5 square units

j. 8 square units

There are many ways to find the areas of these figures. One useful technique involves drawing a rectangle around the figure. This drawing shows a rectangle around figure i. To find the area of the figure, subtract the sum of the areas of the triangles from the area of the rectangle.

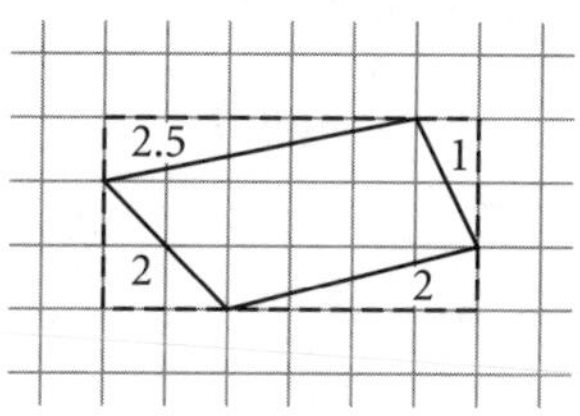

$$\text{Area of figure i} = 3 \cdot 6 - (2.5 + 1 + 2 + 2)$$
$$= 18 - 7.5 = 10.5$$

Steps 2–4 If you know the area of a square, you can find the side length by taking the square root. For example, the square labeled d in Step 1 has an area of 2 square units, so the length of each side is $\sqrt{2}$ units. The square labeled e in Step 1 has an area of 8 square units, so the length of each side is $\sqrt{8}$ units.

Look at the squares in Step 3. The first square has an area of 9 and a side length of 3.

To find the area of the second square, surround it by a square with horizontal and vertical sides as shown here. The area is 10 square units, so the side length is $\sqrt{10}$ units.

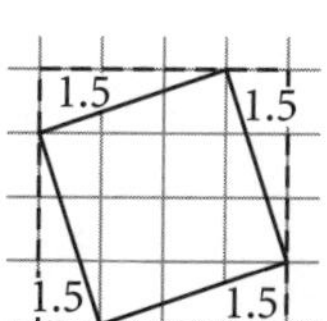

$$\text{Area} = 16 - 4(1.5)$$
$$= 16 - 6$$
$$= 10 \text{ square units}$$

(continued)

Step 4 shows the smallest and largest squares that can be drawn on a 5-by-5 grid. Draw at least five other squares, and find the area and side length of each. Here are three examples.

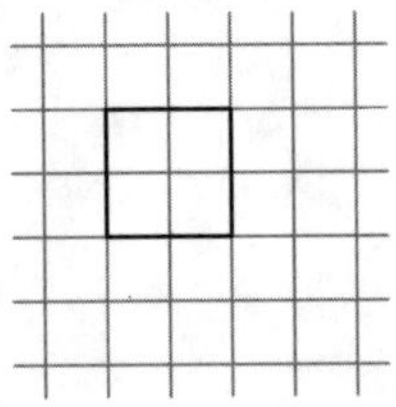

Area = 4 square units
Side length = 2 units

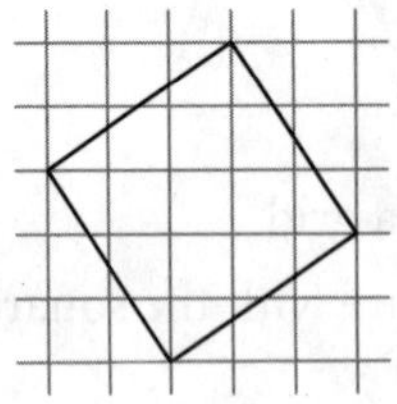

Area = 13 square units
Side length = $\sqrt{13}$ units

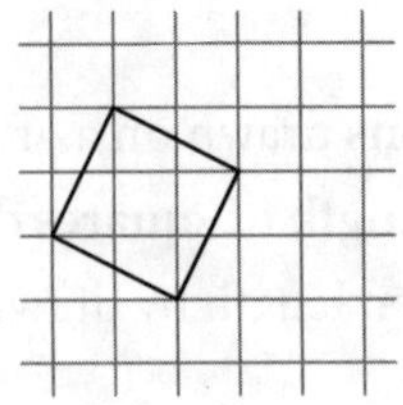

Area = 5 square units
Side length = $\sqrt{5}$ units

Because $\sqrt{10}$ is equal to a decimal whose digits go on forever, it may surprise you that you can draw a segment with a length of exactly $\sqrt{10}$ units. You just need to draw a square with an area of 10 square units; each side will have a length of $\sqrt{10}$ units. The example below shows you how to draw a segment with a length of $\sqrt{17}$ units.

EXAMPLE | Draw a line segment that is exactly $\sqrt{17}$ units long.

▶ ***Solution***

To draw a segment with a length of $\sqrt{17}$ units, first draw a square with an area of 17 square units. Because 17 is not a perfect square, the square will be on a tilt. Start with a larger square, such as a 5-by-5 square. A 5-by-5 square has an area of 25 square units. Try to draw a tilted square inside the 5-by-5 square so that the sum of the areas of the four surrounding triangles is $25 - 17$, or 8.

Here are two ways to draw a tilted square with vertices on a 5-by-5 square.

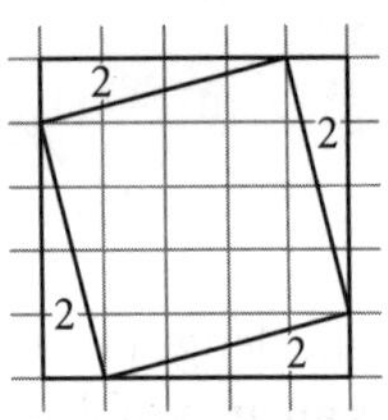

Area of tilted square = $25 - 4(2) = 17$ sq. units

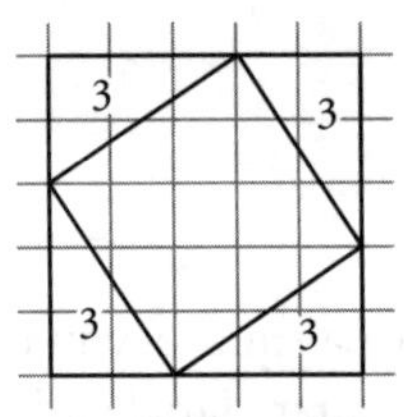

Area of tilted square = $25 - 4(3) = 13$ sq. units

The square on the left has an area of 17 square units, so each of its sides is a segment with a length of $\sqrt{17}$ units.

CONDENSED LESSON

11.4 The Pythagorean Theorem

In this lesson you will

- discover the **Pythagorean theorem**
- use the Pythagorean theorem to find the missing side length of a right triangle

In your book, read the text before the investigation. It explains that the area formula for a triangle is

$$Area = \frac{base \cdot height}{2} \quad \text{or} \quad A = \frac{1}{2}bh$$

Investigation: The Sides of a Right Triangle

The diagram on page 599 of your book shows a right triangle with squares drawn on the sides. Find the area of each square and record the results in a table like the one shown in Step 4. Then, copy each right triangle below, draw a square on each side, and record the areas in your table. Repeat these steps for two right triangles of your own.

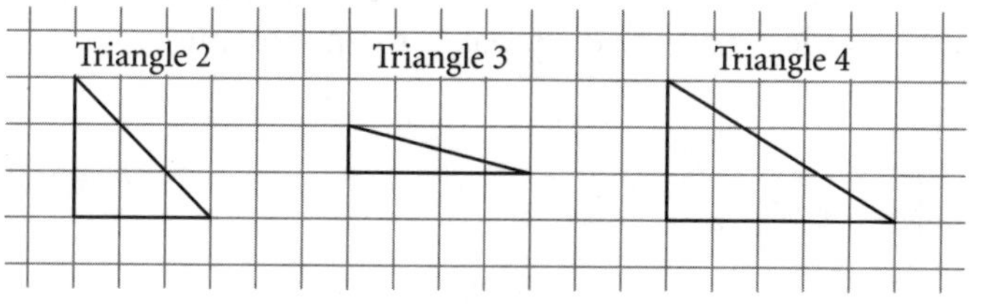

Here are the results for the triangle in your book and the three triangles above.

	Area of square on leg 1	Area of square on leg 2	Area of square on the hypotenuse
Triangle 1	4	16	20
Triangle 2	9	9	18
Triangle 3	1	16	17
Triangle 4	9	25	34

For each right triangle, you should find that the area of the square on the hypotenuse is equal to the sum of the areas of the squares on the legs.

Now, calculate the lengths of the legs and the hypotenuse for each triangle.

	Length of leg 1	Length of leg 2	Length of hypotenuse
Triangle 1	2	4	$\sqrt{20}$
Triangle 2	3	3	$\sqrt{18}$
Triangle 3	1	4	$\sqrt{17}$
Triangle 4	3	5	$\sqrt{34}$

(continued)

For each right triangle, you will find that this rule relates the lengths of the legs to the length of the hypotenuse.

$(\textit{length of leg } 1)^2 + (\textit{length of leg } 2)^2 = (\textit{length of hypotenuse})^2$

The relationship you discovered in the investigation is known as the **Pythagorean theorem.** A *theorem* is a mathematical formula or statement that has been proven to be true. Read the statement of the theorem on page 600 of your book.

The Pythagorean theorem is useful for finding the length of one side of a right triangle when you know the lengths of the other two sides. The example in your book shows how to use the theorem to find the distance from home plate to second base on a baseball diamond. Read that example and then read the example below.

EXAMPLE

The size of a television set or computer monitor is described by giving the length of the diagonal of its screen. The screen on Jackson's 27-inch television has a height of about 16.25 inches. How wide is the screen?

▶ ***Solution***

Here is a sketch of Jackson's television. The drawing shows a right triangle, with the height and width of the screen as legs and the diagonal as the hypotenuse.

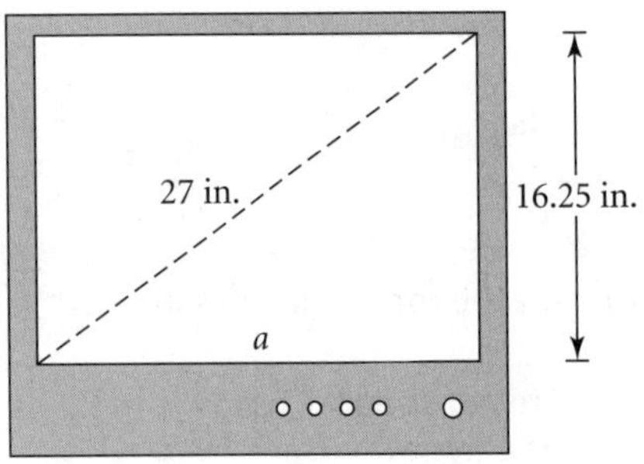

You can use the Pythagorean theorem to find the width of the television.

$a^2 + b^2 = c^2$	The Pythagorean theorem.
$a^2 + 16.25^2 = 27^2$	One leg has a length of 16.25, and the hypotenuse has a length of 27.
$a^2 + 264.0625 = 729$	Calculate the squares.
$a^2 = 464.9375$	Subtract 264.0625 from both sides.
$a = \sqrt{464.9375}$	Take the square root of both sides.
$a \approx 21.56$	Evaluate.

The television screen is about 21.56 inches wide.

CONDENSED LESSON

11.5 Operations with Roots

In this lesson you will

- learn the **rules for rewriting radical expressions**
- apply the rules to find areas of rectangles and check solutions to quadratic equations

Radical expressions can be written in more than one way. In this lesson you'll learn rules for rewriting radical expressions.

Example A in your book shows how you can use the Pythagorean theorem to draw a segment with a length of $\sqrt{13} + \sqrt{13}$. The example indicates that $\sqrt{13} + \sqrt{13}$ is equivalent to $2\sqrt{13}$. Read this example carefully and make sure you understand it. You will use similar methods as you work on the investigation.

Investigation: Radical Expressions

Steps 1–3 On graph paper, draw line segments with the lengths given in Step 1. You may need to draw more than one triangle to create some of the lengths. Here are drawings for parts a, e, and h.

a.

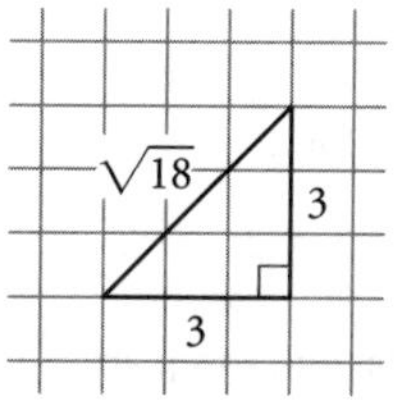

e.

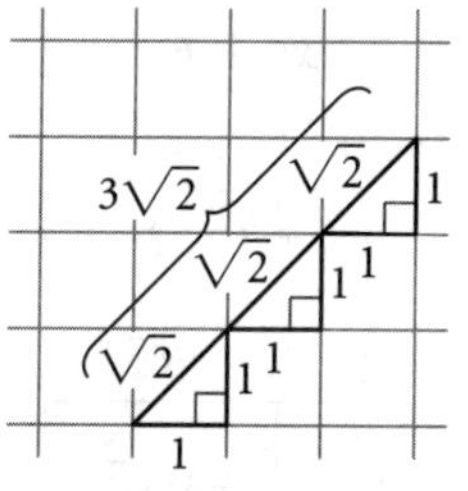

h.

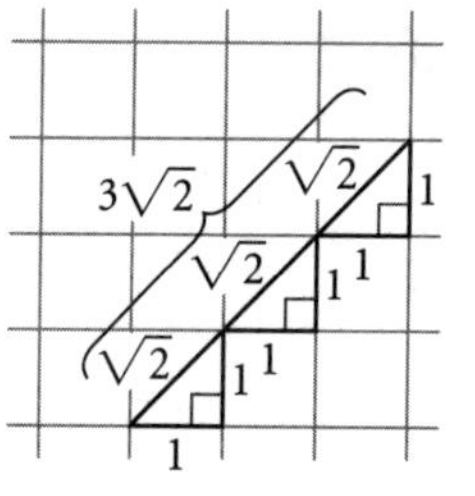

These segments are all the same length. This illustrates that $\sqrt{18}$, $3\sqrt{2}$, and $\sqrt{2} + \sqrt{2} + \sqrt{2}$ are equivalent expressions. (You can verify this by finding decimal approximations with your calculator.)

Your drawings should also show that $\sqrt{40}$, $2\sqrt{10}$, and $\sqrt{10} + \sqrt{10}$ are equivalent and that $\sqrt{20}$, $2\sqrt{5}$, and $\sqrt{5} + \sqrt{5}$ are equivalent.

Steps 4–6 Look at the expressions in Step 4. Try to find another way to write each expression. Substitute positive values for the variables to check that your expression is equivalent to the original. Here are the results.

a. $4\sqrt{x}$ **b.** $\sqrt{xy}$ **c.** $x\sqrt{y}$ or $\sqrt{x^2y}$

d. x or $\sqrt{x^2}$ **e.** $\sqrt{x}$

Here is a summary of some of the things you have learned about rewriting radical expressions.

- You can add or subtract radical expressions with the same number under the square root symbol.
- The square root of a number times the square root of another number is equal to the square root of the product.
- The square root of a number divided by the square root of another number is equal to the square root of the quotient.

(continued)

Use what you have learned to find the area of each rectangle in Step 6. Here are the results for parts b and d.

b. $2\sqrt{7} \cdot 2\sqrt{7} = 2 \cdot 2 \cdot \sqrt{7}\sqrt{7}$

$= 4\sqrt{7 \cdot 7}$

$= 4\sqrt{49}$

$= 4 \cdot 7 = 28$

The area is 28 square inches.

d. $7\sqrt{8}(5 + 6\sqrt{12}) = 35\sqrt{8} + 42\sqrt{96}$

$= 35\sqrt{4 \cdot 2} + 42\sqrt{16 \cdot 6}$

$= 35\sqrt{4} \cdot \sqrt{2} + 42\sqrt{16}\sqrt{6}$

$= 35(2)\sqrt{2} + 42(4)\sqrt{6}$

$= 70\sqrt{2} + 168\sqrt{6}$

The area is about 510.51 square centimeters.

Now, follow along with Example B in your book. Then, read the rules for rewriting radical expressions on page 605. Example C shows you how to use the rules to check a solution to a quadratic equation. Here is another example.

EXAMPLE | Show that $-3 + \sqrt{2}$ is a solution to $x^2 + 6x + 7 = 0$.

▶ ***Solution***

$x^2 + 6x + 7$ Original expression.

$(-3 + \sqrt{2})^2 + 6(-3 + \sqrt{2}) + 7$ Substitute $-3 + \sqrt{2}$ for x.

$(-3 + \sqrt{2})^2 - 18 + 6\sqrt{2} + 7$ Distribute 6 over $-3 + \sqrt{2}$.

Use a rectangular diagram to square $-3 + \sqrt{2}$.

	-3	$\sqrt{2}$
-3	9	$-3\sqrt{2}$
$\sqrt{2}$	$-3\sqrt{2}$	2

$(9 - 3\sqrt{2} - 3\sqrt{2} + 2) - 18 + 6\sqrt{2} + 7$

$9 + 2 - 18 + 7$ Combine terms $-3\sqrt{2}$, $-3\sqrt{2}$, and $6\sqrt{2}$.

0 Add and subtract.

Substituting $(-3 + \sqrt{2})$ for x makes the expression $x^2 + 6x + 7$ equal to 0, so $-3 + \sqrt{2}$ is a solution to the equation $x^2 + 6x + 7 = 0$.

CONDENSED LESSON

11.6 A Distance Formula

In this lesson you will

- discover the **distance formula,** which is used to find the distance between two points
- solve equations involving radical expressions

Example A in your book shows how you can find the distance between two points by plotting them, drawing a right triangle, and applying the Pythagorean theorem. Read this example carefully. In the investigation, you will learn a formula that can be used to find the distance between two points without plotting them.

Investigation: Amusement Park

Steps 1–4 Look at the map of the amusement park on page 612. You can find coordinates for each of the attractions.

a. Acrobats: $(-1, 4)$ **b.** Ball Toss: $(-2, -2)$ **c.** Bumper Cars: $(-4, -3)$

d. Ferris Wheel: $(0, 0)$ **e.** Hall of Mirrors: $(3, 1)$ **f.** Mime Tent: $(3, 3)$

g. Refreshment Stand: $(-5, 2)$ **h.** Roller Coaster: $(-4, 5)$ **i.** Sledge Hammer: $(2, -3)$

Find the exact distance between each pair of attractions listed in Step 2. For attractions that are not on the same horizontal or vertical line, you will need to draw a right triangle and use the Pythagorean theorem. For example, you can use this triangle to find the distance between the Refreshment Stand and the Ball Toss.

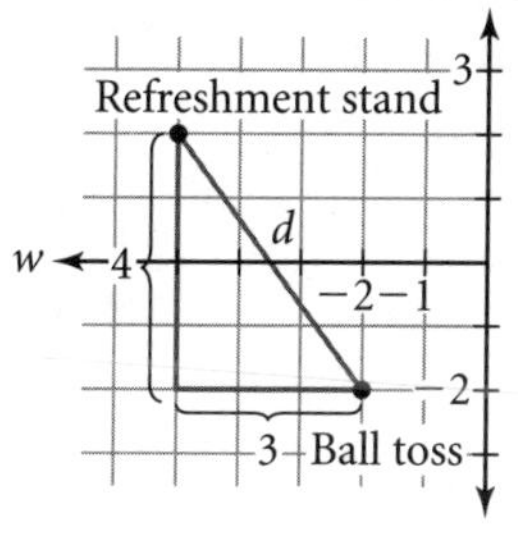

$$3^2 + 4^2 = d^2$$

$$9 + 16 = d^2$$

$$25 = d^2$$

You should get these answers.

a. 6 units **b.** $\sqrt{10}$ units **c.** 2 units

d. 5 units **e.** $\sqrt{85}$ units

The Roller Coaster and the Sledge Hammer are farthest apart. Using the Pythagorean theorem, you can calculate that the distance between them is 10 units. If each unit represents 1 mile, they are 1 mile apart.

Suppose Chris parks his car at $(17, -9)$. Imagine a right triangle with a hypotenuse extending from the Refreshment Stand to Chris's car. The length of the horizontal leg is the difference in the x-coordinates: $17 - (-5) = 22$. The length of the vertical leg is the difference in the y-coordinates: $-9 - 2 = -11$. If d is the length of the hypotenuse, then $d^2 = 22^2 + 11^2$, or 605; so $d = \sqrt{605}$, or about 24.6 units. If each unit is 0.1 mile, this is 2.46 miles.

(continued)

Steps 5–9 Two new attractions are being considered for the park. The first, P_1, will have coordinates (x_1, y_1), and the second, P_2, will have coordinates (x_2, y_2). You can draw a right triangle with legs parallel to the axes and hypotenuse $\overline{P_1P_2}$.

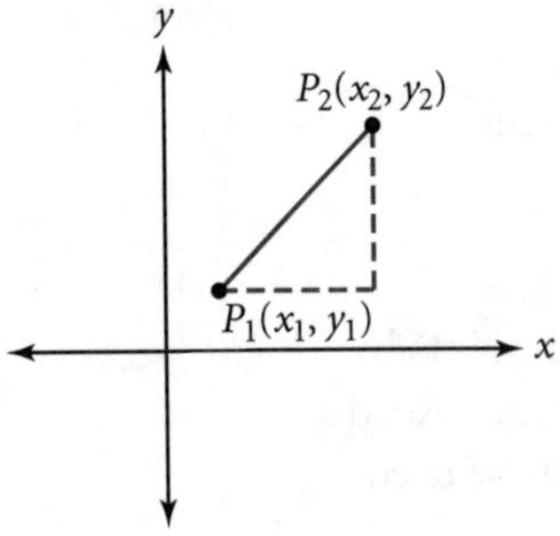

The vertical distance between P_1 and P_2 is $y_2 - y_1$. The horizontal distance between P_1 and P_2 is $x_2 - x_1$. Using the Pythagorean theorem, you know that

$$\text{distance}^2 = (x_2 - x_1)^2 + (y_2 - y_1)^2$$

Taking the square root of both sides, you get the formula

$$\text{distance} = \sqrt{(x_2 - x_1)^2 + (y_2 - y_1)^2}$$

This formula works for any two points. For example, the Bumper Cars have coordinates $(-4, -3)$, and the Mime Tent has coordinates $(3, 3)$. You can use the formula to find the distance between these two attractions.

$$\text{distance} = \sqrt{(3 - (-4))^2 + (3 - (-3))^2} = \sqrt{(7)^2 + (6)^2} = \sqrt{49 + 36} = \sqrt{85}$$

The formula you derived in the investigation is known as the **distance formula.** Read about the formula on page 614 of your book. Also, read Examples B and C. Example C stresses the importance of checking your solutions when you solve a square root equation. Here is another example.

EXAMPLE Solve the equation $\sqrt{2x - 5} = x - 4$.

▸ Solution

$\sqrt{2x - 5} = x - 4$	The original equation.
$(\sqrt{2x - 5})^2 = (x - 4)^2$	Square both sides to undo the square root.
$2x - 5 = x^2 - 8x + 16$	The result of squaring.
$0 = x^2 - 10x + 21$	Add $-2x$ and 5 to both sides.
$x = 3$ or $x = 7$	Use the quadratic formula, a graph, or a table to solve.

Check:

$\sqrt{2(3) - 5} = \sqrt{1} = 1$ and $3 - 4 = -1$, so 3 is *not* a solution.

$\sqrt{2(7) - 5} = \sqrt{9} = 3$ and $7 - 4 = 3$, so 7 is a solution.

CONDENSED LESSON

11.7 Similar Triangles and Trigonometric Functions

In this lesson you will

- solve a proportion to find a missing side length in a pair of similar triangles
- identify the **legs opposite** and **adjacent** to an **acute angle** of a right triangle
- calculate the **sine, cosine,** and **tangent** ratios for the acute angles of a right triangle

You know that *similar polygons* have corresponding sides that are proportional. Example A in your book reviews how to find an unknown side length of similar triangles by solving a proportion. Read and follow along with this example. Then, read the text following the example.

Investigation: Ratio, Ratio, Ratio

Below are four right triangles. In each triangle, the side labeled o is the leg **opposite** angle A and the side labeled a is the leg **adjacent** to angle A. The side labeled h is the hypotenuse.

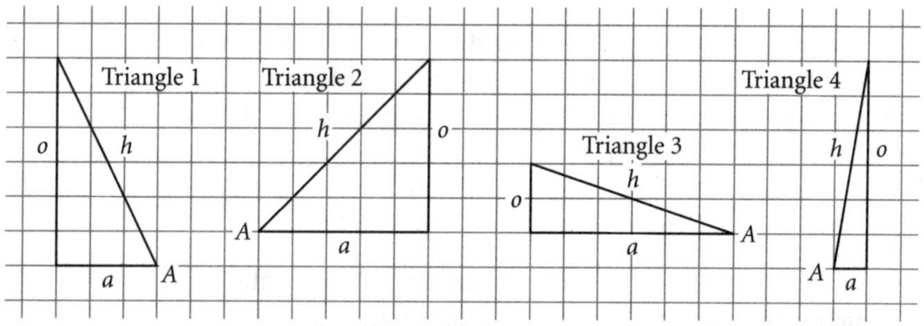

For each triangle, find the measure of angle A, and record it and the side lengths in a table. Draw two or three more right triangles and add their information to the table. Here are the measurements for the triangles above.

	Triangle 1	Triangle 2	Triangle 3	Triangle 4
Measure of angle A	63°	45°	18°	81°
Length of adjacent leg (a)	3	5	6	1
Length of opposite leg (o)	6	5	2	6
Length of hypotenuse (h)	$\sqrt{45} \approx 6.71$	$\sqrt{50} = 7.07$	$\sqrt{40} \approx 6.32$	$\sqrt{37} \approx 6.08$

Calculate the ratios $\frac{o}{h}$, $\frac{a}{h}$, and $\frac{o}{a}$ for each triangle. Here are the results for the same triangles.

	Triangle 1	Triangle 2	Triangle 3	Triangle 4
Measure of angle A	63°	45°	18°	81°
$\frac{o}{h}$	$\frac{6}{\sqrt{45}} \approx 0.89$	$\frac{5}{\sqrt{50}} \approx 0.71$	$\frac{2}{\sqrt{40}} \approx 0.32$	$\frac{6}{\sqrt{37}} \approx 0.99$
$\frac{a}{h}$	$\frac{3}{\sqrt{45}} \approx 0.45$	$\frac{5}{\sqrt{50}} \approx 0.71$	$\frac{6}{\sqrt{40}} \approx 0.95$	$\frac{1}{\sqrt{37}} \approx 0.16$
$\frac{o}{a}$	$\frac{6}{3} = 2$	$\frac{5}{5} = 1$	$\frac{1}{3} = 0.\overline{3}$	$\frac{6}{1} = 6$

(continued)

Now, with your calculator in the degree mode, find the sine, cosine, and tangent of angle A, and record the results in a table. (See **Calculator Note 11A** to learn about these functions on your calculator.) Here are the results for these triangles.

	Triangle 1	Triangle 2	Triangle 3	Triangle 4
Measure of angle A	63°	45°	18°	81°
sine A	0.89	0.71	0.31	0.99
cosine A	0.45	0.71	0.95	0.16
tangent A	1.97	1	0.32	6.31

These results are approximately equal to the ratios in the previous table. If you had measured the angles to the nearest tenth or hundredth of a degree, they would be closer. The sine, cosine, and tangent ratios—abbreviated sin, cos, and tan—are defined as follows.

$$\sin A = \frac{\textit{opposite leg}}{\textit{hypotenuse}} \qquad \cos A = \frac{\textit{adjacent leg}}{\textit{hypotenuse}} \qquad \tan A = \frac{\textit{opposite leg}}{\textit{adjacent leg}}$$

Choose one of the four triangles, and draw a larger right triangle with an acute angle congruent to angle A. Label the congruent angle D. Measure the side lengths, and calculate the sine, cosine, and tangent ratios for angle D. You should find that the ratios are the same as those for the original triangle.

The sine, cosine, and tangent are called *trigonometric functions* and are fundamental to the branch of mathematics called *trigonometry.* Learning to identify the parts of a right triangle and evaluate these functions for particular angle measures is an important problem-solving tool. The box on page 620 reviews what you have learned about these functions.

Example B in your book shows you how to find the sine, cosine, and tangent ratios for particular angles of a triangle. Read Example B and then read the example below.

EXAMPLE

Find these ratios for this triangle.

a. $\cos A$ **b.** $\tan A$ **c.** $\cos B$ **d.** $\sin B$

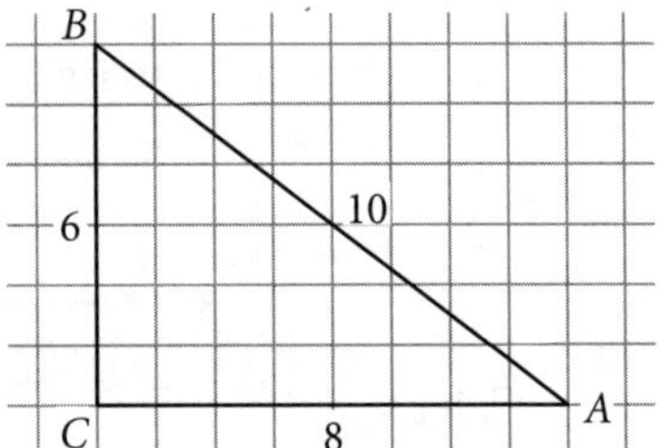

▶ ***Solution***

For angle A, $a = 8$, $o = 6$, and $h = 10$.

a. $\cos A = \frac{a}{h} = \frac{8}{10} = \frac{4}{5}$ **b.** $\tan A = \frac{o}{a} = \frac{6}{8} = \frac{3}{4}$

For angle B, $a = 6$, $o = 8$, and $h = 10$.

a. $\cos B = \frac{a}{h} = \frac{6}{10} = \frac{3}{5}$ **b.** $\sin B = \frac{o}{h} = \frac{8}{10} = \frac{4}{5}$

CONDENSED LESSON

11.8 Trigonometry

In this lesson you will

- use **trigonometric functions** to find side lengths of a right triangle
- use **inverse trigonometric functions** to find angle measures of a right triangle
- use trigonometry to help you interpret a **topographic map**

If you know the measure of on acute angle of a right triangle and the length of one side, you can use trigonometric functions to find the lengths of the other sides. If you know the lengths of the sides, you can use inverse trigonometric functions to find the measures of the angles. These ideas are demonstrated in Examples A and B in your book. Here are two more examples.

EXAMPLE A

Find the value of x in this triangle.

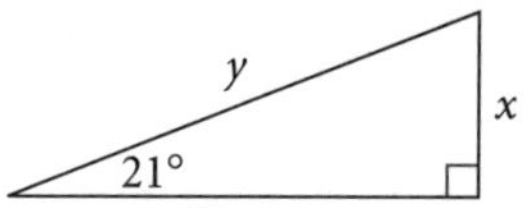

▸ *Solution*

The variable x represents the length of the leg opposite the 21° angle. The length of the adjacent leg is 50 cm.

$\tan A = \frac{o}{a}$	Definition of tangent.
$\tan 21° = \frac{x}{50}$	Substitute 21° for A and 50 for a.
$50 \tan 21° = x$	Multiply both sides by 50.
$19.2 \approx x$	Evaluate the tangent function and multiply.

The length of x is about 19.2 cm. Now, see if you can find the length of y.

EXAMPLE B

Find the measure of angle A.

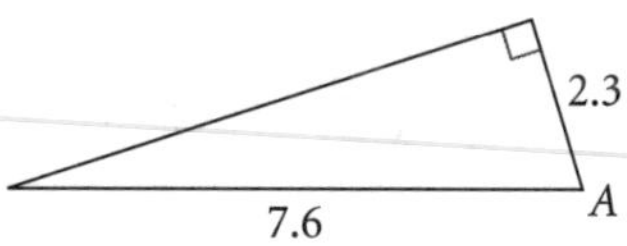

▸ *Solution*

Because you know the length of the leg adjacent to angle A and the length of the hypotenuse, you can find the cosine ratio.

$$\cos A = \frac{2.3}{7.6} \approx 0.303$$

You can find the measure of the angle with the inverse cosine of 0.303.

$$A = \cos^{-1}(0.303) \approx 72°$$

(continued)

Investigation: Reading Topographic Maps

The *topographic map* on page 627 of your book shows the elevation of a hill. On the map, there is a vertical rise of 20 meters between every two rings, or *contour lines.* Study the map until you think you understand it.

The contour lines are farther apart on the west side of the summit than on the east side. This indicates that the rise is more gradual as you approach the summit from the west.

Suppose you hike from the west side, over the summit, and down the east side. The contour lines and the summit divide the hike into 8 sections. The slope of each section is the vertical rise (the change in elevation) divided by the horizontal run (the distance between contour lines). This drawing shows a slope triangle for each section of the hike.

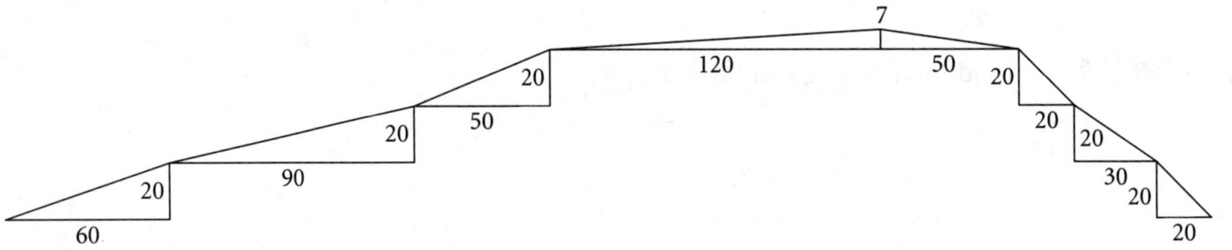

Use the Pythagorean theorem to find the length of the hypotenuse of each slope triangle. This length approximates the actual distance hiked during that section of the hike. Here are the results.

Section 1: 63.2 m Section 2: 92.2 m Section 3: 53.9 m Section 4: 120.2 m

Section 5: 50.5 m Section 6: 28.3 m Section 7: 36.1 m Section 8: 28.3 m

Use the inverse tangent function to find the angle of climb for each section. For example, $\tan^{-1}\left(\frac{20}{65}\right)$ is the angle of climb for section 1.

Section 1: 18° Section 2: 13° Section 3: 22° Section 4: 3°

Section 5: 8° Section 6: 45° Section 7: 34° Section 8: 45°

Measure each angle and compare your results with those above.

Now, read the remainder of the lesson in your book.

Key Curriculum Press
Innovators in Mathematics Education

Comment Form

Please take a moment to provide us with feedback about this book. We are eager to read any comments or suggestions you may have. Once you've filled out this form, simply fold it along the dotted lines and drop it in the mail. We'll pay the postage. Thank you!

Your Name ____________________

School ____________________

School Address ____________________

City/State/Zip ____________________

Phone ____________________ Email ____________________

Book Title ____________________

Please list any comments you have about this book.

Do you have any suggestions for improving the student or teacher material?

To request a catalog, or place an order, call us toll free at 800-995-MATH, or send a fax to 800-541-2242.
For more information, visit Key's website at www.keypress.com.

@4pm

~~10 graphs~~ 8 graphs - label sits: find graph

5 True / False

2 Rotate Graphs

15 total

Fold carefully along this line.

NO POSTAGE NECESSARY IF MAILED IN THE UNITED STATES

BUSINESS REPLY MAIL

FIRST CLASS PERMIT NO. 338 EMERYVILLE, CA

POSTAGE WILL BE PAID BY ADDRESSEE

Key Curriculum Press

Innovators in Mathematics Education

Attn: Editorial Department
1150 65th Street
Emeryville, CA 94608-9740

$y = 3/(x+2)$

Fold carefully along this line.

4 - 4^30 - type lesson schedule

~~5~~

9^30 - 10 shower

10 - 11^45

Key Curriculum Press
Innovators in Mathematics Education

Comment Form

Please take a moment to provide us with feedback about this book. We are eager to read any comments or suggestions you may have. Once you've filled out this form, simply fold it along the dotted lines and drop it in the mail. We'll pay the postage. Thank you!

Your Name ______________________________

School ______________________________

School Address ______________________________

City/State/Zip ______________________________

Phone ______________________ Email ______________________

Book Title ______________________________

Please list any comments you have about this book.

Do you have any suggestions for improving the student or teacher material?

To request a catalog, or place an order, call us toll free at 800-995-MATH, or send a fax to 800-541-2242.
For more information, visit Key's website at www.keypress.com.

Fold carefully along this line.

NO POSTAGE NECESSARY IF MAILED IN THE UNITED STATES

BUSINESS REPLY MAIL

FIRST CLASS PERMIT NO. 338 EMERYVILLE, CA

POSTAGE WILL BE PAID BY ADDRESSEE

Key Curriculum Press
Innovators in Mathematics Education

Attn: Editorial Department
1150 65th Street
Emeryville, CA 94608-9740

Fold carefully along this line.